Learning by Doing with National Instruments Development Boards

Jivan Shrikrishna Parab
Ingrid Anne Nazareth
Rajendra S. Gad
Gourish Naik

CRC Press is an imprint of the
Taylor & Francis Group, an **informa** business

CRC Press
Taylor & Francis Group
6000 Broken Sound Parkway NW, Suite 300,
Boca Raton, FL 33487-2742

and by

CRC Press
2 Park Square, Milton Park, Abingdon, Oxon OX14 4RN

CRC Press is an imprint of Taylor & Francis Group, an Informa business

International Standard Book Number-13: 978-1-138-33833-3 (Hardback)

Visit the Taylor & Francis Web site at
http://www.taylorandfrancis.com

and the CRC Press Web site at
http://www.crcpress.com

Learning by Doing with National Instruments Development Boards

Contents

Foreword

Traditional electronic instruments are rapidly being replaced by virtual instrumentation (VI) due to its inflexibility and high cost. The simplest VI consists of a personal computer loaded with application software, low cost plug-in boards, and driver software. Laboratory Virtual Instrument Engineering Workbench (LabVIEW) from National Instruments (NI) is a popular system design platform and development environment for a visual programming language which is commonly used in VI. SPEEDY 33, myRIO, and NIELVIS are the three most popular development boards from NI which can be used in conjunction with LabVIEW. The frustrating truth for the beginners is the general lack of text books that covers this graphic tool and development boards. This book provides the beginners not only with a pedagogic introduction to these VI tools of NI but also a "hands on experience" that no textbooks can provide.

In fact, this is the fourth book from these authors with a practical hands-on approach. Their earlier books entitled *Exploring C for Microcontrollers: A hands on Approach* [1], *Practical Aspects of Embedded System Design using Microcontrollers* [2], and *Hands on Experience on Altera FPGA Development boards* [3] published by Springer Netherlands are very informative, pedagogic and stimulates self-learning with a practical approach. I have found these texts in several top world university libraries. The authors have adopted the familiar easy pedagogic approach of "Learning by Doing" in this book just as they have done in their previous three books.

All the four authors have a good expertise in embedded and biomedical instrumentation research and development work for several years. During my interaction with them in Goa University, I realized that they have exemplary hands-on experience in embedded systems. This book is the product of their expertise, experience, and passion in VI. It offers an in-depth, yet practical, explanation of the various elements of the popular development boards from NI using LabVIEW.

I am confident that the readers will enjoy the simple language and step-by-step description of LabVIEW programming and using SPEEDY 33, myRIO and NIELVIS boards with ample examples. I strongly recommend this book for beginners, intermediate programmers, and electronics, electrical and instrumentation engineers or any individual who is strongly inclined to take up his or her career in virtual instrumentation graphical programming.

Professor A. Srinivasan
Department of Physics, Indian Institute of Technology Guwahati
Guwahati, India

References

1. www.worldcat.org/title/hands-on-experience-with-altera-fpga-development-boards/oclc/1012344097
2. www.worldcat.org/title/practical-aspects-of-embedded-system-design-using-microcontrollers/oclc/1012557935&referer=brief_results
3. www.worldcat.org/title/exploring-c-for-microcontrollers-a-hands-on-approach/oclc/123114201&referer=brief_results

Preface

LabVIEW (Laboratory Virtual Instrumentation Engineering Workbench) is a platform and development environment for a visual programming language from National Instruments. The graphical language is named "G." Originally released for the Apple Macintosh in 1986, LabVIEW is commonly used for data acquisition, instrument control and industrial automation on a various platforms including Microsoft Windows, various flavors of UNIX, Linux and Mac OS X.

LabVIEW integrates the creation of user interfaces (termed front panels) into the development cycle. LabVIEW programs-subroutines are termed virtual instruments (VIs). Each VI has three components: a block diagram, a front panel and a connector panel. The last is used to represent the VI in the block diagrams of other, calling VIs. The front panel is built using controls and indicators. Controls are inputs; they allow a user to supply information to the VI. Indicators are outputs; they indicate, or display, the results based on the inputs given to the VI. The back panel, which is a block diagram, contains the graphical source code. All of the objects placed on the front panel will appear on the back panel as terminals. The back panel also contains structures and functions which perform operations on controls and supply data to indicators. The structures and functions are found on the Functions palette and can be placed on the back panel. Collectively controls, indicators, structures and functions will be referred to as nodes. Nodes are connected to one another using wires, e.g. two controls and an indicator can be wired to the addition function so that the indicator displays the sum of the two controls. Thus a virtual instrument can be run as either a program, with the front panel serving as a user interface, or, when dropped as a node onto the block diagram, the front panel defines the inputs and outputs for the node through the connector pane. This implies each VI can be easily tested before being embedded as a subroutine into a larger program.

The graphical approach also allows nonprogrammers to build programs by dragging and dropping virtual representations of lab equipment with which they are already familiar. The LabVIEW programming environment, with the included examples and documentation, makes it simple to create small applications. This is a benefit on one side, but there is also a certain danger of underestimating the expertise needed for high-quality G programming. For complex algorithms or large-scale code, it is important that a programmer possess an extensive knowledge of the special LabVIEW syntax and the topology of its memory management. The most advanced LabVIEW development systems offer the ability to build stand-alone applications. Furthermore, it is possible to create distributed applications, which communicate by a client–server model, and are thus easier to implement due to the inherently parallel nature of G.

One benefit of LabVIEW over other development environments is the extensive support for accessing instrumentation hardware. Drivers and abstraction layers for many different types of instruments and buses are included or are available for inclusion. These present themselves as graphical nodes. The abstraction layers offer standard software interfaces to communicate with hardware devices. The provided driver interfaces save program development time.

Chapter 1 includes a brief glimpse of LabVIEW, virtual instrumentation, the relationship between VI and LabVIEW, basics of LabVIEW which includes a user interface and the front panel, with controls and indicators. Controls are knobs, push buttons, dials and other input devices. Indicators are graphs, LEDs and other displays. This chapter also includes programming with LabVIEW which consists of arrays, loops, clusters, structures, plotting of data, etc. The chapter will also feature how to install the software and license setup.

Chapter 2 describes the various board configuration setups. There are three different configurations, namely NI SPEEDY-33, NI ELVIS and myRIO. This chapter will explain in detail the different configurations and the differences between each of them.

Chapter 3 gives hands-on experience of performing experiments on the NI SPEEDY-33 board. This chapter includes various basic experiments like interfacing the board with LEDs, switches, keypad, digital filter design, modulation, audio signal processing (echo, reveberation) and creating digital music.

Chapter 4 includes various experiments performed using the NI ELVIS board. These include experiments like 4-Bit Adder, Traffic Light Control, Digital Thermometer and Hearing Aid.

Chapter 5 describes basic experiments performed on myRIO. The basic experiments include keypad interfacing, pushbutton switch interface, LED interfacing, seven-segment LED display, UART, SPI and I2C interface.

Authors

Dr. Jivan Shrikrishna Parab is assistant professor in the Department of Electronics at Goa University, India. He completed his PhD from the same university with the thesis titled "Development of Novel Embedded DSP Architecture for Non-Invasive Glucose Analysis." He received his MSc (2005) and (2003) BSc in electronics degrees from Goa University. He has co-authored three books, published by Springer: *Practical Aspects of Embedded System Design Using Microcontrollers*, *Exploring C for Microcontrollers: A Hands-on Approach* and *Hands-on Experience with Altera FPGA Development Boards*. He has published several papers in national and international level journals and conferences. Presently he is member of faculty board and library committee of Goa University. Recently he has been awarded with the Visvesaraya Young Faculty award of Rs. 38 lakhs by the government of India.

Dr. Ingrid Anne Nazareth, born in Sharjah, UAE, is currently working as assistant professor in the Department of Electronics at Goa University, India. She completed her PhD from the same university with the thesis titled "Estimation of Human Blood Cholesterol." She completed her Masters in Electronics having secured first place and is an awardee of the "IV SERC School in Physics Gold Medal." She was appointed a project fellow in the project "Design of Hyperspectral Smart Sensors Using Soft-Core Processors and IP Cores." She was also a visiting faculty at the Goa University. Her research interest is in the field of biomedical electronics. She has attended a number of national symposiums and conferences where she has presented her research work.

Dr. Rajendra S. Gad is professor of Electronics at Goa University. He received BSc (Physics) and MSc (Electronics) degrees from Goa University in 1995 and 1997 respectively. He worked for his PhD in areas of non-invasive measurements to understand problems of human body glucose measurement in 2009. His group was judged winner in Indiato design LC3 processor at Mentor Graphics University Design Contest 2010. He established MOU with AlTERA Inc. USA under a university program to develop an FPGA SoC laboratory. His areas of interest are micro-UAVs, real time system verifications, smart sensors systems, signal processing and networks. He has co-authored two books in the area of embedded systems published by Springer and Lambart Publishing.

Prof. Gourish Naik (Dean Faculty of Natural Science, Head Department of Electronics) obtained his PhD from the Indian Institute of Science, Bangalore (1987) and served the institute as research associate in the areas of Opto-electronics and Communication until 1993. For the last 25 years, he has been associated with the Goa University Electronics Program. He is the founding head of the University Instrumentation Center and established fiber optic LAN and wireless communication networks at Goa University. He was also the coordinator of DEITI (an educational broadcast studio supported by Indian Space Research). His other commitments are regulating the digitization center at Goa University to support the various digital repository projects like DIGITAP (Digital Repository for Fighter Aircrafts Documentation) of the Indian Navy, the Million Book project of Ministry of Information Technology and Antarctica Study Center (NCAOR). He has to his credit around 50 odd research papers published in international journals and has presented research works at various national and international forums. He has delivered several key note addresses and has been invited to talks at various institutes and also authored four books on embedded systems and allied areas published by Springer, Lambert, etc. He was a member of Goa State Rural Development Authority and also advisor for Directorate of Education. He is governing body member of the engineering college of Goa and also a member of the faculty board of Goa University. Presently he is head of the Electronics Department at Goa University.

CHAPTER ONE

A Glimpse of LabVIEW

1.1 What is Virtual Instrumentation

Virtual instrumentation (VI) is a multifaceted field to combine software and hardware technologies so as to create instruments both sophisticated and flexible for monitoring as well as control applications. There are quite a few descriptions of a Virtual Instrument.

1. A Virtual Instrument as described by Santori is "an instrument whose capabilities and general functions are decided by the software" [1].
2. Goldberg says "A Virtual Instrument is made up of various general-purpose computers, some software, a few dedicated subunits and a little technology" [2,3].
3. These descriptions capture the fundamental ideas of VI and virtual concepts overall which are affordable: "any computer can replicate another, if it is loaded with software to simulate another computer" [4].

1.2 History of VI

The history of VI is differentiated by continuous enhancement of scalability and flexibility of the measurement equipment. Beginning with the first vendor-defined manual-controlled electrical instruments, the field of instrumentation has progressed toward current sophisticated measuring equipment which is user-defined and controlled by a computer.

It has progressed through the following stages:

1. Analog measurements
2. Data acquisition and data processing
3. Digital processing
4. Distributed VI

The first stage represents analog devices such as EEG or oscilloscope recording systems which were completely sealed and dedicated systems inclusive of sensors, power suppliers, displays and translators [5]. Manual settings were required to display results on several counters, CRT displays, gauges or paper. Additionally, the user had to copy data onto a data sheet or a notebook physically, since copying data was not possible. Subsequently, everything had to be done physically since performing automated tests or complex procedures was rather impossible.

The second stage was launched in the 1950s since the industrial control field insisted on the same, wherein rudiment control systems with integrators, rate detectors and relays were incorporated in the instruments. This led to the formation of proportional-integral-derivative (PID) control systems that permitted flexibility of automation of quite a few phases of measuring processes and test procedures [2,3]. Digital processing of data was possible since signals were digitalized in the instruments.

In stage three, measuring instruments were computer based, which included interfaces that allowed communication connecting the computer and the instrument. In the 1960s, Hewlett-Packard (HP), previously known as HPIB, started the relationship which was started by the General-Purpose Interface Bus (GPIB) for the purpose of control of instruments by the HP computers. Originally computers were made use of as off-line instruments which were further used to process data after recording on a disk [6]. Computers became faster for real-time measurements as the capabilities and the speed of the computers increased. Then manufacturers incorporated the necessary hardware and software required by the instruments for the specific applications. The most important advantages of PCs are standardization and availability due to the low price in the market. In 1986 National Instruments launched LabVIEW 1.0 [1], which had Graphical User Interfaces (GUI) making it user friendly and affordable.

The fourth stage came into being with the development of global and local networks of computers. With the development of computerized instruments, physical distribution of VI components was made possible to offer medical information and services from afar with the help of network and telecommunication technologies. Infrastructure for allocated VI includes cellular networks, private networks and the internet, where the interface amidst the components could be balanced for performance and price [2,3].

1.3 LabVIEW and VI

Computers are combined by VI with hardware and software. VI are used by engineers to carry out the power of PC technology to design, control and test applications and flexible software to make accurate digital and analog measurements. Scientists and engineers can produce user-defined systems to meet the specific application needs. Manufacturers with automated processes, for instance manufacturing or chemical plants, use VI with the aim of improving reliability, system productivity, stability, optimization and safety. VI is a computer software that any user could employ to build up a computerized measurement and test system in order to control from a desktop, an external hardware device and also for displaying measurement or test data accumulated by any external device on panels like a computer screen. It is also used to computerize systems for organizing processes based on collected and processed data. The control function of the front panel is duplicated on the computer interface. Applications range from simple experiments in the laboratory to large automations. VI as shown in Figure 1.1 uses synchronizing platforms, modular I/O and productive software.

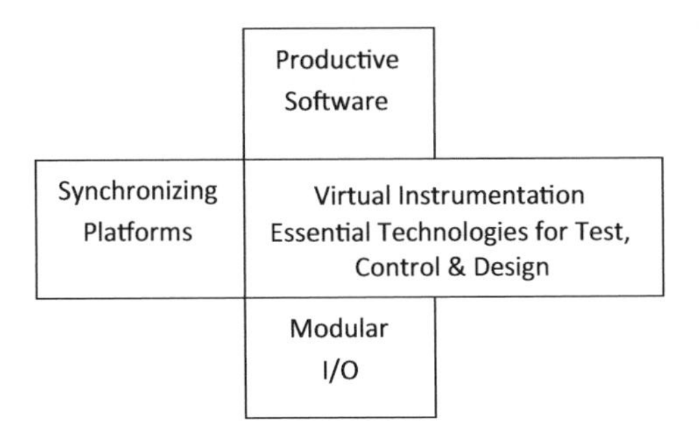

FIGURE 1.1 VI combines synchronizing platforms, modular I/O and productive software.

The first VI element uses Synchronizing Platforms to ensure that VI takes advantage of the data transfer technologies and the latest computer capabilities. The element delivers VI on a long-standing technology base which ranges with the huge investments made in buses, processors and much more. The second VI component is a Modular I/O, designed to be combined rapidly in any quantity or order to make sure that VI can either control or monitor any developing feature. Scientists and engineers and can rapidly access functions by using ingenious software drivers for Modular I/O during synchronized operation. Finally, in order to speed up development, NI LabVIEW, a premier VI graphical developed environment, uses graphical or symbolic representations. The software uses symbols to represent various functions. VI is termed as a computer having low cost driver software and hardware and user friendly application software which performs the functions of instruments together. VI is also known as simulated physical instruments. With VI, scientists and engineers lower their design costs, design higher quality products and decrease development time.

VI is essential since it is flexible. It carries on instrumentation with the swift adaptability needed for today's delivery, concept, development, process and product design. With these instruments, scientists and engineers can keep up with upcoming demands with the latest technology. For example, a mobile phone now has Internet Browsing, Bluetooth Networking, MP3 Player and Camera, although it was originally used to call and send SMS.

VI consists of hardware and software and is required to fulfill the control or measurement task. Further, scientists and engineers can customize sharing, storage, presentation functionality, analysis and acquisition using powerful and productive software. A multiple purpose VI can be manufactured by using a data acquisition card or board of which the benefits include size, ease of programming, cost and flexibility. VI software has many layers such as application software, data and test management software, and control and measurement services software. The application software is the basic development environment used to build applications including software like LabVIEW, VI Logger, Signal Express, Measurement Studio and LabWindows/CVI. Above this layer is the data and test management software which incorporates the functionality of the previous layer and offers data management. Control and measurement services software is like the I/O driver S/W layer which connects the VI hardware and software for control and measurement. GUI played an important role in developing VI. The main developments to help VI were increasing steady networking platforms, developing bus standards, evolution of GUI and developing low cost computers with sufficient computing power. The resource intensive techniques were developed by the improvement of processor computing power.

History of LabVIEW (**Lab**oratory **V**irtual **I**nstrument **E**ngineering **W**orkbench)

1983: NI minimized the time required to program instrumentation systems

1986: LabVIEW1.0 released for Macintosh as an interpreted package

1990: LabVIEW2.0 released as a compiled package. Stable graphical environment

1992: LabVIEW2.5 released for Windows and Sun

1993: LabVIEW3.0 released for Windows, Sun and Macintosh

1994: LabVIEW3.0.1 released for Windows NT

1994: LabVIEW3.1 released

1995: LabVIEW3.1.1 released with application builder

1996: LabVIEW released for Linux

1996: LabVIEW4.0 released with debugging tools and editing for instrumentation

1997: LabVIEW4.1 released

1998: LabVIEW5.0 and 5.1 released (control framework, dynamic programming and built-in Web server)

1999: LabVIEWRT released

2000: LabVIEW6.0 released (intuitive and easy programming interface)

2001: LabVIEW6.1 released (remote Web control event-oriented programming of LabVIEW)

2003: LabVIEW7.0 released (Assistants and Express utilities available)

2003: LabVIEWPDA released

2003: LabVIEWFPGA released

2004: LabVIEW7.1 released

2005: LabVIEW released embedded module

2005: LabVIEW8.0 released

2006: LabVIEW8.2.0 released

2007: LabVIEW8.2.1 released

2007: LabVIEW8.5 released

2008: LabVIEW8.6 released

2008: LabVIEW8.6.1 released

2009: LabVIEW2009 9.0.0 released

2010: LabVIEW2009 9.0.1 released

2010: LabVIEW2010 10.0.0 released

2010: LabVIEW2010f2 10.0.0 released

2011: LabVIEW2010SP1 10.0.1 released

2011: LabVIEW LEGO MINDSTORMS released

2011: LabVIEW2011 11.0.0 released

2012: LabVIEW2011SP1 11.0.1 released

2012: LabVIEW2012 12.0.0 released

2012: LabVIEW2012SP1 12.0.1 released

2013: LabVIEW2013 13.0.0 released

2014: LabVIEW2013SP1 13.0.1 released

2014: LabVIEW2014 14.0 released

2015: LabVIEW2014SP1 14.0.1 released

2015: LabVIEW2015 15.0 released

2016: LabVIEW2015SP1 15.0.1 released

2016: LabVIEW2016 16.0.0 released

2017: LabVIEW2017 17.0 released

2018: LabVIEW2017SP1 17.0.1 released

2018: LabVIEW2018 18.0 released

2018: LabVIEW2018SP1 18.0.1 released

2019: LabVIEW2019 19.0 released

LabVIEW uses a graphical environment with a graphical language known as G programming which uses block diagrams graphically to be compiled into machine code and removes syntactical details. It is flexible, it can create the type of VI required and the user can view and modify control or data inputs easily. The programs in LabVIEW are known as VIs since its operation and appearance imitates physical instruments. It is designed to simplify collection of data and it's analysis and display options. LabVIEW consists of a complete set of functions and VIs to acquire, analyze, store and display data in addition to a troubleshoot code.

The advantages of LabVIEW are as follows:

1. Graphical User Interface (GUI)
2. Modular design and hierarchical design
3. Professional development tools
4. Preserves investment and reduces cost
5. Connectivity and instrument control
6. Distributed development
7. Rapid development with express technology
8. Simple application distribution
9. Object-oriented design
10. Drag-and-drop built-in functions
11. Multiple high level development tools
12. Multi platforms
13. Algorithm design
14. Flexibility and scalability
15. Open environment
16. Visualization capabilities
17. Compiled language for fast execution
18. Target management

1.4 LabVIEW Basics

Components of LabVIEW

Software Environment Three steps are needed in the software environment to create an application:

- To design an user interface
- To draw a graphical code
- To run the program

VI has three major components—icon/connector pane, block diagram and front panel. To use a VI as a subVI within the block diagram of a second VI, it is important that it should contain a connector and an icon. The connector is a set of terminals used as indicators and controls of the VI. The icon is present in the top right corner of the block diagram and the front panel. It contains images and texts which are graphically represented in the VI. The two LabVIEW windows consist of a block diagram (graphical code, connections and terminals) and a front panel (indicators and controls). The block diagram consists of a graphical source code. The front panel is the user interface of the VI. A code is built by using functions in graphical form in order to control objects in the front panel. In LabVIEW, a front panel is built with indicators and controls. Indicators consist of LEDs, graphs and other displays. Controls consist of dials, push buttons, knobs and many other input devices. After building the user interface, a code using structures and VIs can be added to control the objects of the front panel, which is contained in the block diagram.

Palettes: LabVIEW has three graphical and floating palettes, namely functions, controls and tools, which help to build and run VIs.

Block Diagram—Functions Palette: The block diagram is created using the Functions palette. The Functions palette is shown in Figure 1.2. It consists of functions, sub VIs and constants present in the block diagram. In order to use the Functions palette, one must right click in the block diagram to display it, or it can also be viewed by selecting the View menu and then clicking on Functions Palette. The functions and VIs on the palette depend on the view selected. The functions and VIs are present on the sub-palettes based on the function and VI type.

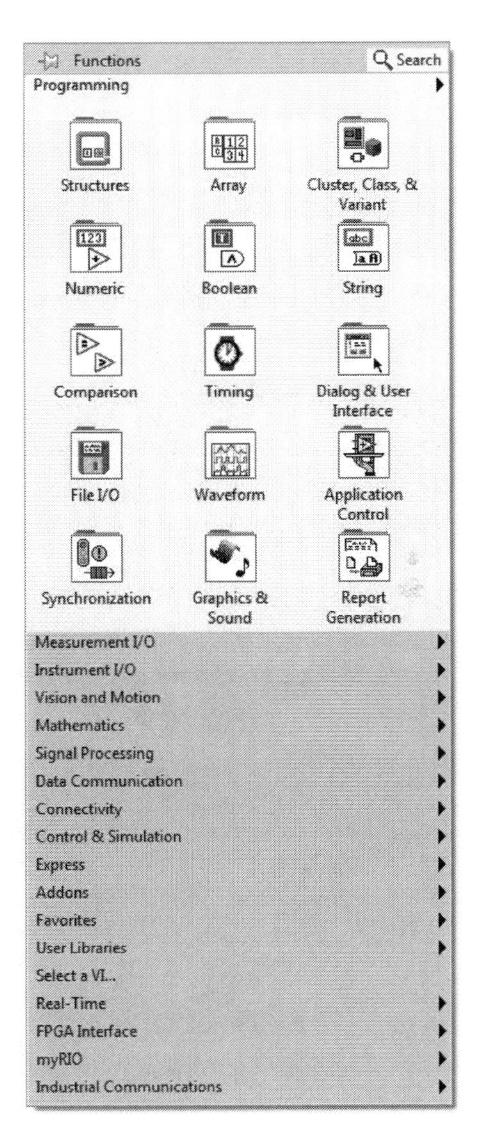

FIGURE 1.2 Functions Palette.

Front Panel—Controls Palette: The front panel is created using the Controls palette. The Controls palette is shown in Figure 1.3. It consists of indicators and controls used to create the front panel. In order to use the Controls palette, one must right click in the front panel to display it, or it can also be viewed by selecting the View menu and then clicking on Controls Palette. VI can also be created by the objects in the sub-palettes present in the Controls palette. When a sub-palette icon is selected, the sub-palette below it is displayed. Click and drag a particular object to the front panel in order to display it.

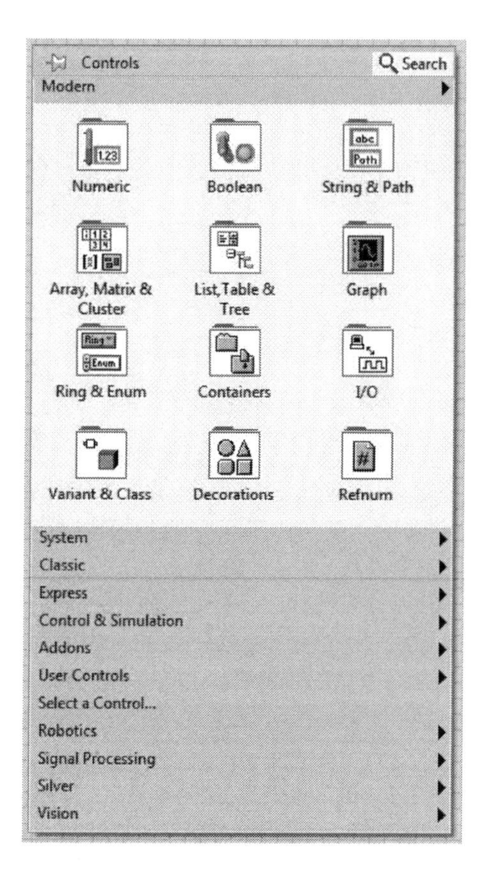

FIGURE 1.3 Controls Palette.

Tools Palette: The Tools palette is shown in Figure 1.4. It is present on the block diagram and the front panel. The tools present on the Tools palette are used to create, modify and then debug VIs. The tool is a unique operating mode of the cursor corresponding to the icon of the tool present in the Tools palette. The tools are used to modify and operate the objects of the block diagram and the front panel. The tools needed can be selected directly on the Tools palette or by selecting the View menu and then clicking on the Tools Palette. The Shift Key + right click can also be used to display the Tools palette.

FIGURE 1.4 Tools Palette.

The Front Panel

When a new VI or existing VI is opened, the front panel appears, which is a user interactive interface. The front panel is shown in Figure 1.5. It includes indicators, graphs, push buttons and knobs. Indicators are used as output devices and to display data that the block diagram receives or generates. Controls are used as inputs to send data and simulate input devices in the block diagram. The front panel emulates the control panel of conventional instruments and also produces test panels or represents operation and control of processes.

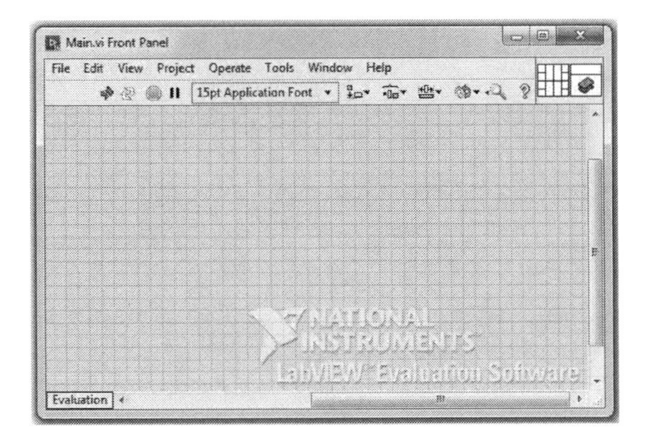

FIGURE 1.5 VI LabVIEW Front Panel.

The most powerful feature that LabVIEW offers to scientists and engineers is the graphical programming to design VI by creating a GUI on the computer's screen to

- Operate the program
- Control the selected hardware
- Analyze the acquired data
- Display the results

Block Diagram

The program of the front panel is accompanied by the block diagram. The objects of the front panel appear as icons in the block diagram and the components are wired together. Thereafter, the front panel is built and codes are added using functions which are represented graphically in the block diagram wherein the objects of the front panel are controlled. The block diagram is shown in Figure 1.6. It contains a graphical source code made up of wires, terminal and node. The block diagram consists of components which maybe lower-level VIs, program execution control structures, constants and built-in functions. The objects have to be connected with wires to ensure the flow of data. Measurement code can be created by using functions, standard VIs and express VIs. The objects of the block diagram include constants, wires, structures, subVIs and terminals. LabVIEW is a powerful and easy tool for receiving, analyzing and displaying data. Terminals are similar to constants and parameters in text-based programming languages. They are the entry and exit ports which exchange information between the block diagram and front panel. In order to change the icon view, right click in the block diagram and then select the View As icon.

FIGURE 1.6 VI LabVIEW Block Diagram.

1.5 Programming with LabVIEW

Modular Programming

Programs are easier to write, to read, to maintain and to debug when divided further into small subprograms. This is known as modular programming. There are many benefits of modular programming. Firstly, it makes writing programs easier because the individual components can be separately written and tested. Secondly, it makes reading easier as long codes can be replaced with simple ones, hence programs are easier to modify. Thirdly, reusing individual components in other programs is made easier. If a program is written to accept data from the computer keyboard to calculate the average and standard deviation present in another function, it can be reused in a separate program which can calculate the average and standard deviation. Although the simple structure of the program works well for a simple example, it is counter-productive for longer programs, resulting in a lack of clarity and a slowing down in the modification and maintenance of code. Every well-defined program task should be placed in the program unit of its own, known as sub-routines. The main program is simply an outline or driver that triggers execution of the units of the program that perform the tasks. The modular programming approach is shown in Figure 1.7.

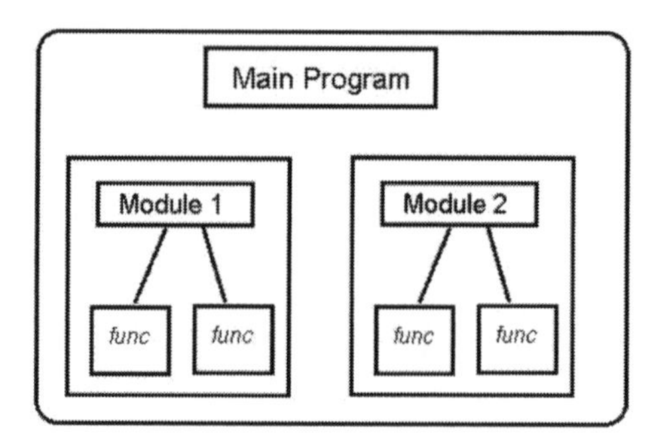

FIGURE 1.7 Modular Programming.

LabVIEW's power lies in the VI's hierarchical character. It can be used on another block diagram of a VI after a VI is created. However, the number of layers is not limited. Modular programming assists in managing changes and quickly debugs the block diagram. Any change to a component or subVI or module has less impact on other subVIs. Modularity enhances the reusability and readability of the VIs. Any operation can be performed frequently as VIs are created, especially using modules or loops.

Arrays and Loops

Arrays An array is a simple data structure consisting of a group of similar elements of a particular data type. An array contains a sequence of data elements of the same data type and size placed in adjacent memory locations which can be referred individually. Therefore arrays store many values below the same name. Each element is accessed by the individual position in the array given by an index also known as a subscript. The index often uses a successive range of integers. The arrays can be one, two or multi dimensional. Arrays are suitable to store data from graphs, waveforms, data collected from multiple sensors or data generated in loops. LabVIEW uses data structures and arrays for this purpose. In LabVIEW, elements of a similar type are grouped in arrays, consisting of dimensions and elements. A dimension is the depth, height or length of an array whereas an element is data that form an array.

An array indicator or control can be created by placing an array icon by selecting Controls on the front panel, then Modern and finally selecting Arrays, Matrix & Clusters. An object must be inserted into the array icon or else it will appear blank as shown in Figure 1.8. This can now be expanded in the horizontal or vertical direction. In the block diagram window, the array can also be inserted by right clicking and then selecting Functions, then Programming and finally selecting Arrays.

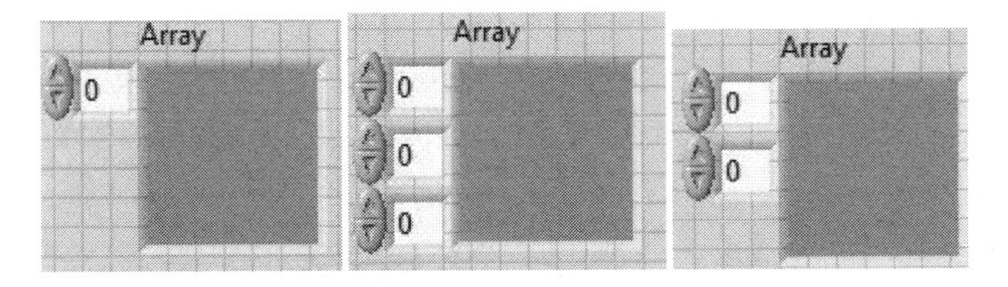

FIGURE 1.8 Blank Array.

The array type will be shown in Figure 1.9 left diagram if it is not a numeric indicator and is shown in Figure 1.9 right diagram with a data type [DBL] on it if it is a numeric indicator.

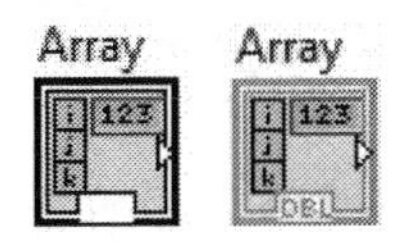

FIGURE 1.9 Array.

A one dimensional array can be converted into a two dimensional array by right clicking on the index display of a particular array and then selecting Add Dimension. A multi dimensional array can be created by further right clicking on the index display on a particular array and then selecting Add Dimension. An array can be uninitialized if it is left blank or initialized if data is entered in it. In order to insert elements in a particular row or column, right click on the array and then select Data Operations and then Insert Row Before or Column Before on the front panel. In order to delete a particular row or column, right click on the array and then select Data Operations and then Delete Row or Delete Column on the front panel. The above are shown in Figure 1.10.

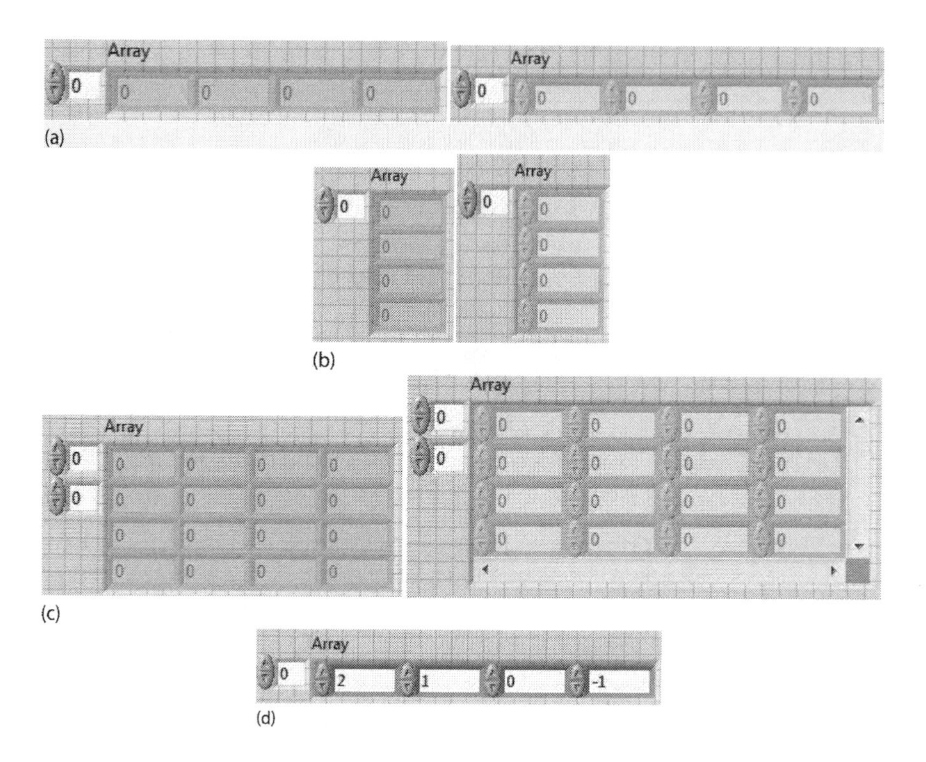

FIGURE 1.10 Types of arrays. (a) Horizontal blank array with indicator and control elements. (b) Vertical blank array with indicator and control elements. (c) Two dimensional array with indicator and control elements. (d) Initialized array.

Loops Loops and case statements based on text programming languages in graphical programming are represented as structures. Loops and repetitions are used each time to execute an action with changes in the details. LabVIEW consists of While and For Loops which are used to control operations repeatedly. The block diagram Structures can repeat blocks of code and execute code in a specific or a conditional order. Besides While and For Loops, LabVIEW also includes Formula Node, Event, Flat Sequence, Stacked Sequence and Case structures.

- *While Loop*: It is similar to a Repeat-Until or Do Loop in other programming languages. A subdiagram is executed until a condition is reached. The While Loop is performed at least once or until the final condition is reached. In the Block Diagram click on Function, then Programming and select Structures and finally While Loop. Now click and drag from the top left corner to the right bottom corner to draw the While Loop as shown in Figure 1.11. It consists of the Conditional and Iteration Terminals. The loop execution is controlled by the Conditional Terminal while the number of completed iterations is controlled by the Iteration Terminal.

FIGURE 1.11 While Loop.

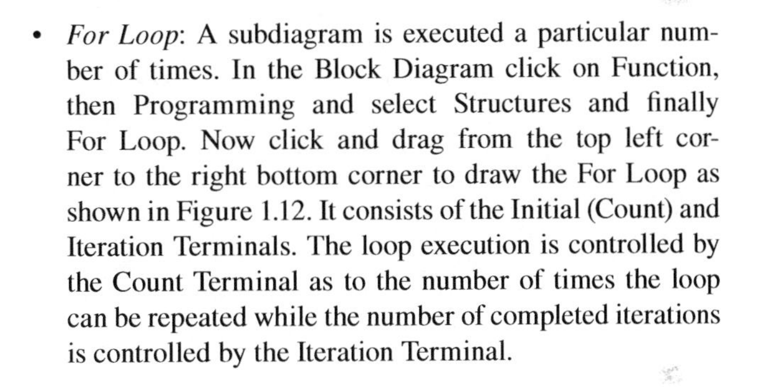

- *For Loop*: A subdiagram is executed a particular number of times. In the Block Diagram click on Function, then Programming and select Structures and finally For Loop. Now click and drag from the top left corner to the right bottom corner to draw the For Loop as shown in Figure 1.12. It consists of the Initial (Count) and Iteration Terminals. The loop execution is controlled by the Count Terminal as to the number of times the loop can be repeated while the number of completed iterations is controlled by the Iteration Terminal.

FIGURE 1.12 For Loop.

- *Shift Register*: In case previous data is needed in the following iterations, either a Shift Register or a Feedback Node is required. A Shift Register is similar to static variables and looks like a pair of terminals on either side of the loop border with arrows in opposite directions opposite to each other. The right side arrow faces up and stores data at the end of an iteration which is transferred to the next iteration. If the Shift Register is not initialized, then While Loops generate default data. For Loops generate default data if an uninitialized array is wired to a For Loop with auto indexing enabled or if 0 is wired to the Count Terminal. In order to transfer values, Shift Registers are used along with While and For Loops from one loop to another. A Shift Register can be added to any loop by right clicking on the border of the loop and selecting Add Shift Register.

- *Feedback Node*: This is generated when the output node is directly connected to its own input node. It appears automatically if the output node or a group of nodes is wired to the input nodes or a group of nodes of a While or For Loop. When an iteration of a loop is completed, it stores data and the value is sent to the next iteration. Long wires can be avoided by using Feedback Nodes. The direction in which the data flows through the wire is indicated by the Feedback Node arrow and changes direction if the flow of data changes.

Structures and Clusters

Structures A structure is a graphical representation of case and loop statement based on text programming languages. When decisions should be taken in a program, then case or if-else statements are used. One of the ways is the select function under the function palette which selects two values depending on the Boolean input. Structures can be used to repeat blocks and execute code in a specific or conditional order. The terminals in the structures connect nodes in a block diagram, which can be executed automatically when input data is present and can supply data after execution is completed to the output wires. Every structure has a resizable and distinctive border to enclose a section of the block diagram which executes to the rules of the structure. A subdiagram is a part of a block diagram within the structure border. Tunnels are the terminals which feed in and out of the structures and are also a connection point. The following structures besides While and For Loops are located on the *Structures* palette controlling the execution process of a block diagram:

- *Case Structures*: Multiple subdiagrams which execute one at a time depend on the input value are known as case structures. They can have two or more cases subdiagrams, but only one is visible. First a case structure is placed on the block diagram. In order to determine which case needs to be executed, an input value is wired to the selector terminal, which can be Boolean, integer, enumerated or string type value. Error can be handled by wiring the selector terminal to an error cluster. The objects are placed inside the case structures in order to execute them and create subdiagrams, which can be duplicated. If a Boolean data type is selected, then it can have a true and false case. If it is an integer, enumerated or string type value, then there are many cases. The labeling tool is used to enter a range of values as well as a single value at the top of a case structure. The name of the selector value present on the top border of the case structure is the case to be selected as shown in Figure 1.13. There are increment and decrement arrows present on either of the case label to select a different case accordingly. A down arrow is also present next to the increment arrow which leads to a drop down menu to select a particular case.

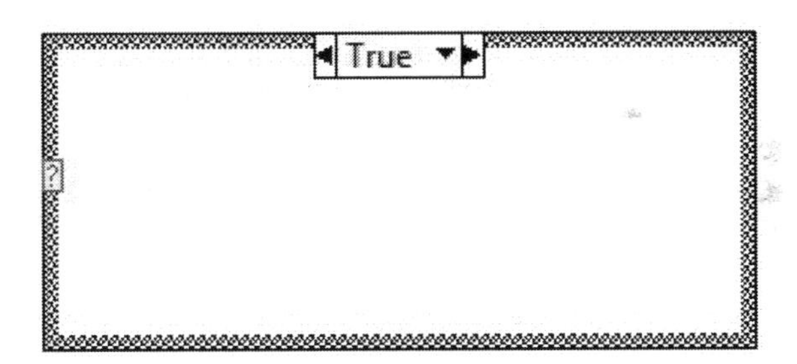

FIGURE 1.13 Case structure.

- *Sequence Structures*: If subdiagrams are executed sequentially, then it is known as sequence structure. The node execution depends on the data dependency and is used when flow through or natural data dependency do not exist. The two kinds of sequence structures are flat and stacked sequence structures.
    - *Flat Sequence Structure*: Here all the frames are displayed at a time and are executed from left to right up to the last frame only when the data is available

which is wired to the frame. As the frame executes, the data leaves the frame until it finishes. Adding or deleting frames resizes the structure automatically. In order to add a flat sequence structure, right click on the block diagram, Functions, Programming, Structures and then on Flat Sequence. Now click and drag from the top left corner to the right bottom corner to draw the flat sequence structure as shown in Figure 1.14.

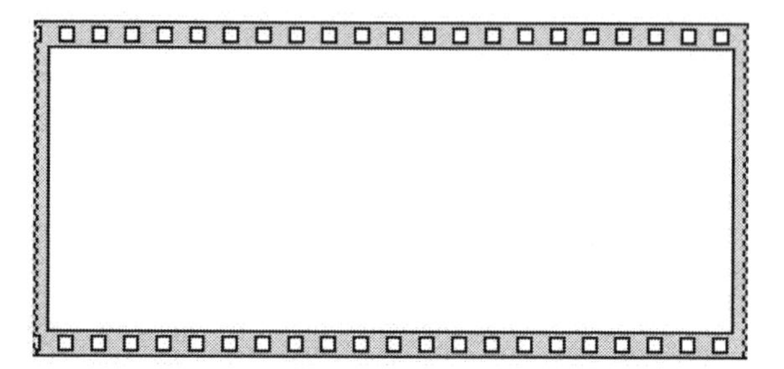

FIGURE 1.14 Flat Sequence structure.

- *Stacked Sequence Structure*: Here, only a single frame is seen at a time which executes from frame 0 onwards up to when the last one is executed and then the data is returned. It is used to save space in the block diagram. In order to convert from a stacked to a flat sequence structure, right click on the stacked sequence structure and then select replace with flat sequence as shown in Figure 1.15. At the top of the stacked sequence, the current frame number and the range of frames is seen.

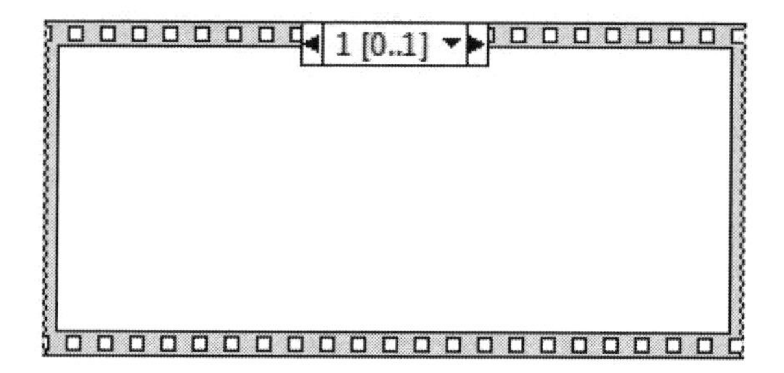

FIGURE 1.15 Stacked Sequence structure.

- *Timed Structures*: A subdiagram is executed in a particular order with time delays and time bounds. The timed structure has a resizable and distinctive border that encloses a part of the block diagram and is known as subdiagram, which follows a set of rules to be executed. In order to give configuration data and receive timing and error information, there are input and output nodes which feed data in and out of the structure. The following structures control the execution of a block diagram with time delays and time bounds.

  - *Timed Loop*: It is executed until a condition is satisfied.
  - *Timed Sequence*: Many subdiagrams are executed in sequence.
  - *Timed Loop with Frames*: Many subdiagrams are executed in sequence until a condition is satisfied.

Clusters A cluster is a group of mixed data element type. For example, a LabVIEW error cluster can consist of a string, numeric value and Boolean value, i.e. a source, code and status. It is similar to a struct or record in programming languages. When many data elements are present in the block diagram, it leads to a lot of wire clutter which can be avoided by bundling it into clusters. The connector pane, which can have a maximum of 28 terminals, can be reduced by passing many values to a single terminal. If indicators or controls need to be passed from one VI to another, they are grouped in a cluster which is assigned to a terminal present on the connector pane.

Clusters can be selected by right clicking on the front panel and then selecting Controls, All Controls and Arrays & Clusters. Click and drag the cluster onto the front panel, after which any data element or object can be dragged into the cluster. The cluster shell can be resized by dragging the sides of the cluster in that particular direction. The cluster appears grey if it is blank in the front panel as shown in Figure 1.16. If a numeric indicator is placed inside a cluster, it looks like Figure 1.17a. If a numeric control is placed inside the cluster, it looks like Figure 1.17b. The figure shows two numeric indicators/constants placed in a cluster. The data type is defined when a particular object is placed in the constant shell.

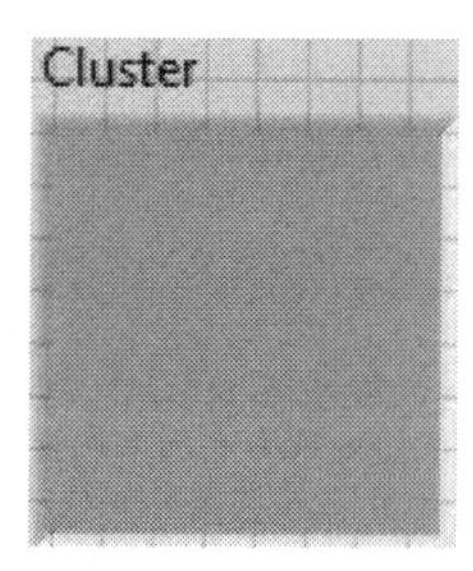

FIGURE 1.16 Blank Cluster.

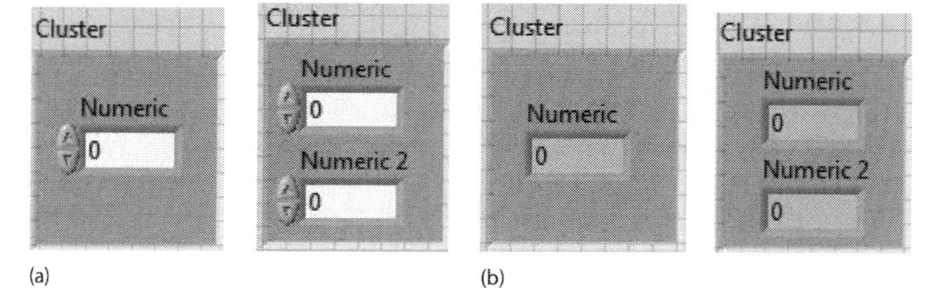

FIGURE 1.17 (a, b) Initialized Clusters.

The different types of cluster operations are unbundled, bundled, unbundle by name and bundle by name. In order to create and manipulate a cluster, cluster functions are used. Many tasks can be performed. Data elements can be extracted from a cluster followed by adding them to the cluster. A cluster can be broken into data elements. The "Bundle Function" is used to assemble each component into an individual cluster and elements can be replaced in the present order, whereas the "Unbundled Function" divides the cluster into separate components. The "Bundle by Name" function is used to operate on a fewer number of components, which are referred by name and not position, whereas the "Unbundle by Name" function returns the specific name from the cluster elements.

Plotting Data

The main aspect of LabVIEW programming is data which is displayed graphically. Cluster and array knowledge is significant for graphical operations. Data is collected in an array and then plotted to get a waveform in VI. 2D XY displays are required everywhere. Graphs and charts display data plots graphically. Graphs are used to plot arrays which are pre-generated in a traditional fashion without keeping the previous generated data. Charts plot data and add new data to the old one such that the current value can be seen with respect to the previous one as new data is generated.

The different types of charts and graphs are as follows:

XY graphs: Data which is received is displayed at a non-steady rate and also for multivalued functions.

Waveform graphs and charts: Data which is received is displayed at a steady rate.

Windows 3D graphs: 3D data is displayed in an ActiveX object on a 3D plot on the front panel.

Digital waveform graphs: Data is displayed as groups or pulses of digital lines.

Intensity graphs and charts: 3D data is displayed using color, on a 2D plot, to display 3D values.

A Cartesian general purpose object which plots many functions like waveforms with a variable time base or with circular shapes is an XY graph. It can be evenly or unevenly sampled, which can be used to display Nichols planes, Nyquist planes, *Z* planes and *S* planes. Labels and lines on the planes are the same color as the Cartesian lines, but the plane label font cannot be modified. Three data types, i.e. a cluster containing two arrays, an array of points (each point is a cluster containing two different values) and an array of complex data (real on the x-axis and imaginary on the y-axis), are accepted by XY graphs for single plot. In multiple plots of XY graphs, there are three data types, i.e. an array of plots (each plot is a cluster containing two different arrays), an array of complex data (real on the x-axis and imaginary on the y-axis) and array of a cluster of an array of points (each point is a cluster containing two different values).

Waveforms charts and graphs in LabVIEW display data at a steady rate. One or more plots in waveform graphs are displayed with measurements that are evenly sampled with a number of points. The points that are evenly distributed are plotted on the x-axis whereas the single valued functions are plotted on the y-axis. Many data types are accepted for plotting a waveform graph of a single plot. Waveform data types, clusters as well as arrays of values, are accepted by graphs for a single plot. In multiple plots of waveform graphs, 2D arrays of values, cluster containing two different values and an array of clusters are accepted.

Instrument and Motion control

Instrument control The instrument control consists of instruments having a serial or a General Purpose Interface Bus (GPIB) interface. LabVIEW is used to acquire and control data instrument drivers, API, VISA and Instrument I/O Assistant. The most frequently used instrument interfaces include PXI modular instruments, serial and GPIB instruments. Other instruments include NI-CAN, parallel port, Ethernet, USB, motion control and image acquisition. Communication protocols are needed to comprehend properties of an instrument to control them when PCs are used. The issues are cables required, communication protocols used, electrical properties, software drivers and types of connectors.

GPIB is a standard, 8 bit parallel and digital communication interface with a data rate of greater than 1Mbps with a 3-wire handshake between controllers and instruments. The bus usually supports around fourteen instruments and one system controller. The GPIB protocol classifies devices such as listeners, talkers or controllers to decide which device controls the bus. Every device consists of an exclusive GPIB address that ranges from 0 to 30. Controllers instruct listeners to listen and interpret data from GPIB and address the talker to talk and put the data on the GPIB. Several devices can listen whereas only one device can talk at a time.

Motion control The various constituents of motion control system are shown in Figure 1.18.

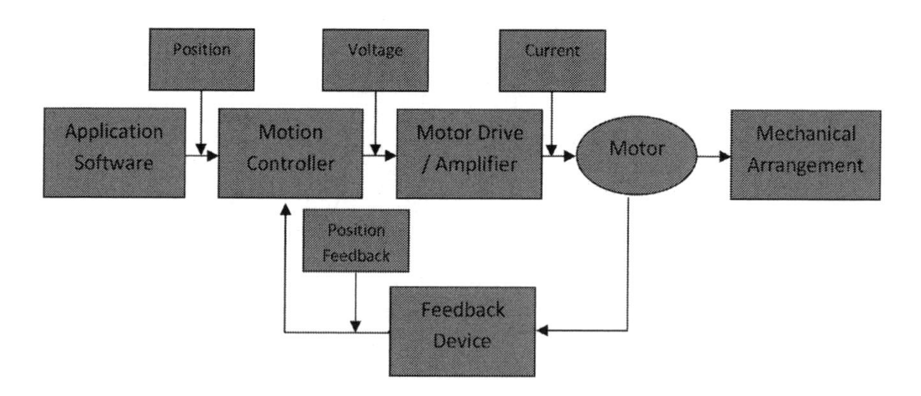

FIGURE 1.18 Constituents of a Motion Control System.

1. *Application software*: It is used to control motion control and target positions.
2. *Motion controller*: It creates a path for the motors to pursue, but outputs a direction and step pulses in the case of stepper motors or a ±10 V signal in the case of servomotors.
3. *Motor drive/Amplifier*: Commands are taken from the controller and the required current is generated to turn or drive the motor.
4. *Motor*: Here electrical energy is converted into mechanical energy and the required torque is produced to shift to the final target position.
5. *Feedback device*: It is important for servomotors. The motor positions are sensed, results are reported and the loop is closed toward the motion controller.
6. *Mechanical arrangement*: Torque is provided to mechanics, which include special actuators, robotic arms and linear slides.

Every move trajectory is calculated by the motion controller, which is an important task, so it is calculated on the DSP present on the board. It is calculated to decide the torque command properly in order to cause motion. The PID control loop is also closed by the motion controller. It closes by itself since a large deterministic level and consistent operation is required. It also monitors the emergency stops and limits for safe operation. Making every operation take place in real time or on the board, in order to make a motion control system, guarantees determinism, reliability, safety and stability.

The three major families of DSP motion controllers offered by NI include the 733x series, 734x series and 735x series. The 733x series offer 4-axis stepper motor control and functions needed for applications that include single and vector motion. The 734x series offers four axes of the servo as well as stepper motor control and features like electronic gearing and contouring. The 735x series offer eight axes of the servo and stepper motor control besides extra I/O and features like 4MHz periodic breakpoints and sine commutation for brushless motors.

The different move types are single axis, point to point (P2P), coordinated multi axis or vector motion, blended, contoured motion and electronic gearing. The position, velocity and acceleration are required by the single axis to which it needs to shift. The vector motion moves from P2P in 2D or

3D space. The final position in the three axes, vector velocity and acceleration are required by the vector motion and finds applications in automated or scanning microscopy. When a blend fuses, two move together, which causes them to move as one. This is known as blended motion and is used where continuous motion is required between two moves. In contoured motion, a position buffer is supplied to create a smooth spline or path through it, and every position is passed unlike blended motion. The simulation of the motion in electronic gearing is done without real gears between two mating gears, and a gear ratio is supplied between a master and slave axis, ADC channel or encoder.

1.6 Installation and Licensing Setup

Step 1: Go to the DVD drive of the computer and double click on setup.

Step 2: Click on Install NI LabVIEW 2016 myRIO Software Bundle.

Step 3: Exit all applications and virus scanning applications and click on Next as shown in Figure 1.19.

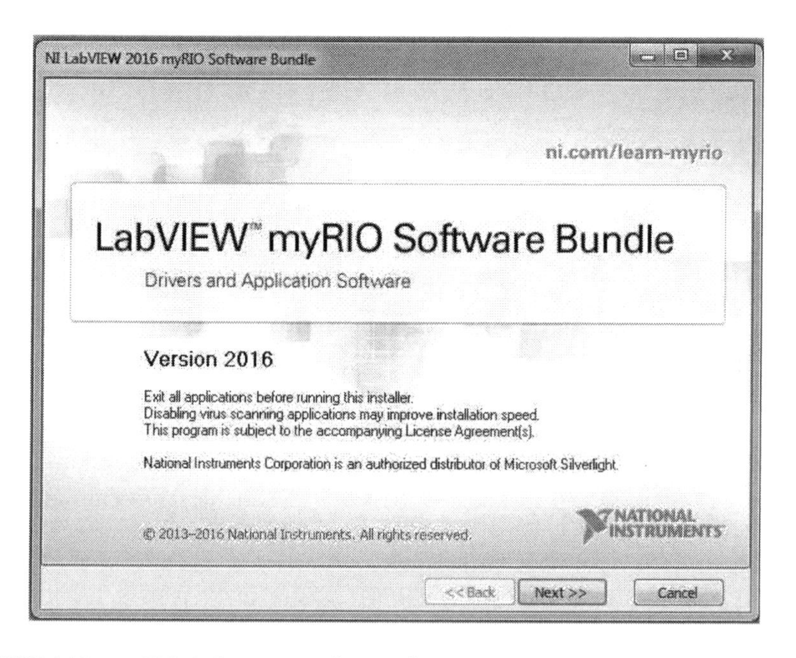

FIGURE 1.19 myRIO Software Bundle Install.

Step 4: Select install for products necessary to be installed and click on Next as shown in Figure 1.20.

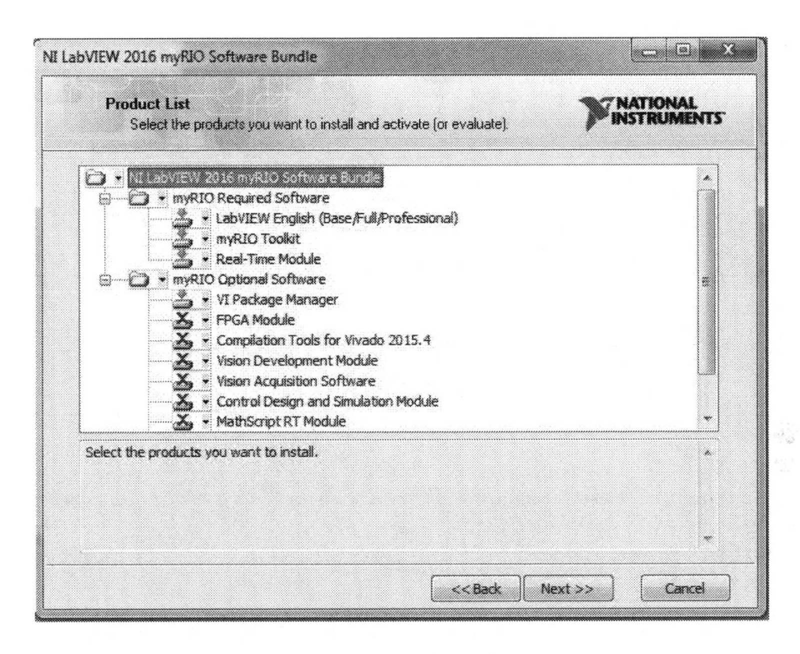

FIGURE 1.20 Product selection for install.

Step 5: Check the box if updates have to be installed and click on Next as shown in Figure 1.21.

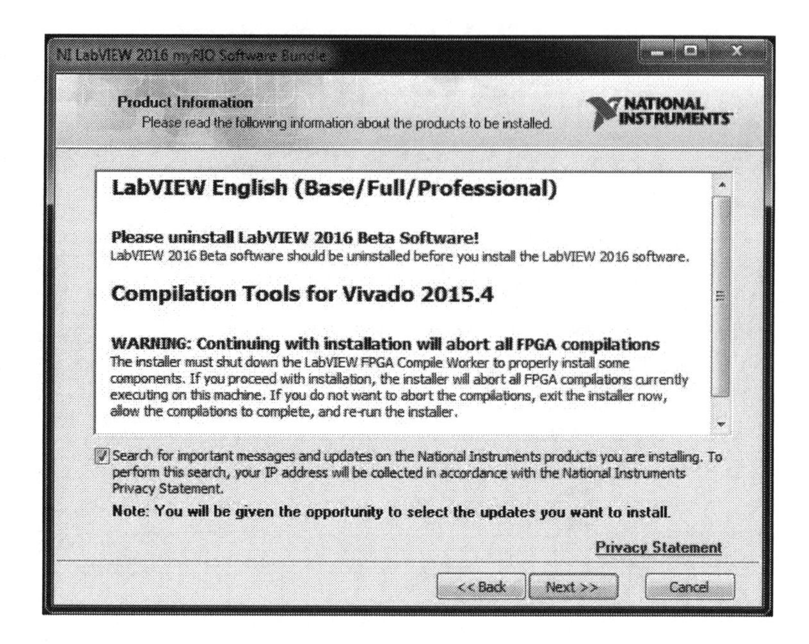

FIGURE 1.21 Check box for updates.

Step 6: Enter the full name and organization as well as the serial number and click on Next. If the Serial Number field is left blank, then the software will run in evaluation mode.

Step 7: Now browse the destination directory for the installation of the NI software and LabVIEW.

Step 8: After selecting the destination directory, click on Next as shown in Figure 1.22.

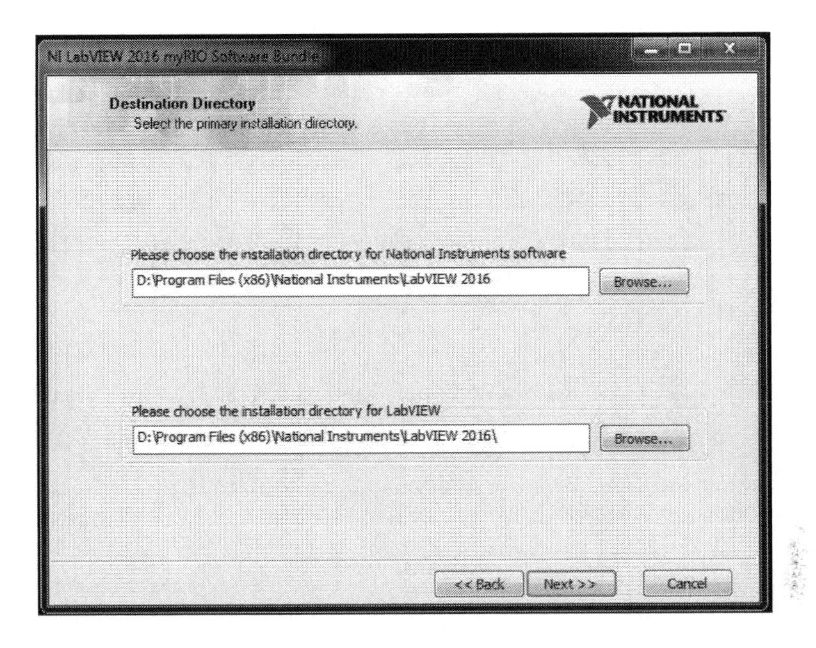

FIGURE 1.22 Selection of Destination directory.

Step 9: Select the "I accept the above 6 License Agreement(s)" and click on Next.

Step 10: Select the "I accept the above 2 License Agreement(s)" and click on Next.

Step 11: Review the list of items to be installed and click on Next as shown in Figure 1.23.

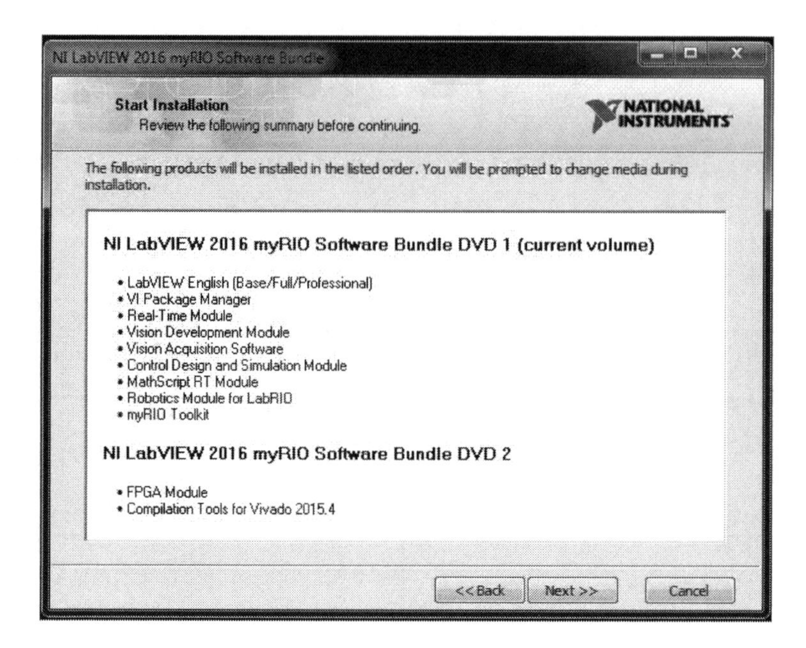

FIGURE 1.23 List of selected items for install.

Step 12: Now the different items will be installed.

Step 13: If the installation from first DVD is completed then scan the particular drive and click on Rescan Drive.

Step 14: All the products are successfully installed. Now select Run License Manager to activate the products and click on Next as shown in Figure 1.24.

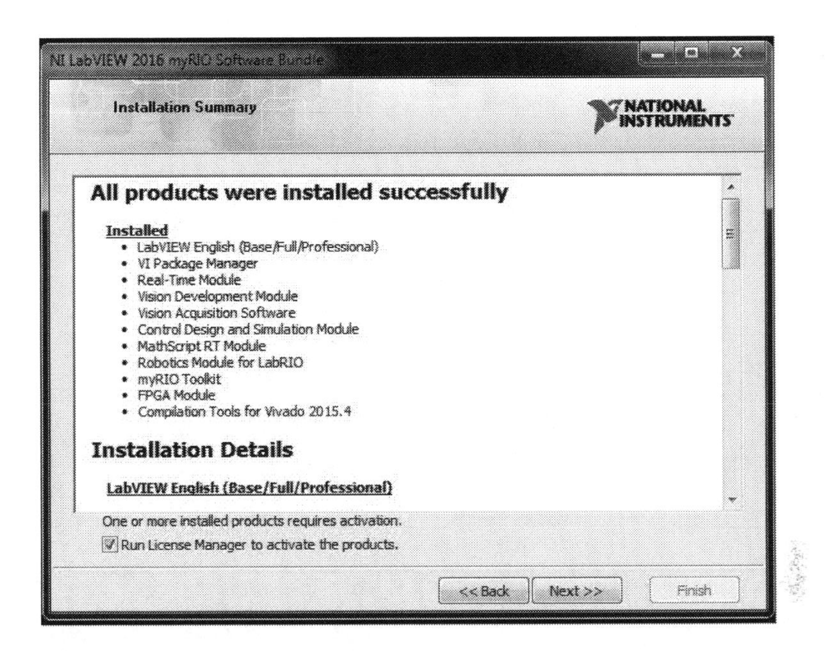

FIGURE 1.24 License manager to activate.

Step 15: Choose Automatically activate through a secure Internet connection or select apply 20 character activation code acquired for the computer and click on Next.

Step 16: Enter the current activation codes and click on Next.

Step 17: Enter the serial numbers for the products needed to be activated.

Step 18: Create a new account and finally login to the NI user account.

1.7 LabVIEW Features

LabVIEW provides several features and tools ranging from interactive assistants to user-defned interfaces which are configurable. It is different from others by its graphical, General purpose programming language (known as G) along with a compiler integrated in it, a linker and debugging tools. It also supports Intuitive Graphical Programming & Interactive Debugging Tools. Apart from these common features of LabVIEW, recently a few more features have been added such as

- *LabVIEW NXG FPGA Module*—Supports USRP (Universal Software Radio Peripheral) and Kintex-7 FlexRIO targets and provides new features for Quick FPGA development and debugging.
- *LabVIEW NXG Web Module*—It has events and properties for dynamic web applications, support for integrating JavaScript Libraries into WebVIs and also helps in hosting WebVIs in a simple and secure way.
- *Development Environment Enhancements*

References

1. M. Santori. An instrument that isn't really, *IEEE Spectrum*, 27(8):36–39, 1990.
2. H. Goldberg. What is virtual instrumentation? *IEEE Instrumentation and Measurement Magazine*, 3(4):10–13, 2000.
3. K. Goldberg. Introduction: The unique phenomenon of a distance. In K. Goldberg (ed.), *The Robot in the Garden: Telerobotics and Telepistemology in the Age of the Internet*, MIT Press, 2000.
4. P.J. Denning. Origin of virtual machines and other virtualities. *IEEE Annals of the History of Computing*, 23(3):73, 2001.
5. L.A. Geddes and L.E. Baker. *Principles of Applied Biomedical Instrumentation*, 3rd edition. John Wiley & Sons, 1989.
6. F. Nebeker. Golden accomplishments in biomedical engineering. *IEEE Engineering in Medicine and Biology Magazine*, 21(3):17–47, 2002.

CHAPTER TWO

Board features and configuration setup

2.1 NI SPEEDY-33

Features

The National Instruments Signal Processing Engineering Educational Device for Youth (NI SPEEDY-33) depicting the Texas Instruments DSP TMS320VC33 is a programmable, high-performance, low cost and self-contained product for specialized DSP applications. It possesses a quick, easy-to-use DSP as well as important features for many applications. The board has an easy-to-use, easy-to-learn software programming tool and maintains the fast establishment of standalone DSP products. The board supports two I/O analog channels which are sampled at 48 KHz, on-chip memory of 136 KB, eight onboard flash memory of 512 K each, eight digital I/O lines as an input port and eight digital LEDs as an output port [1]. Figure 2.1 shows the top view layout of NI SPEEDY-33.

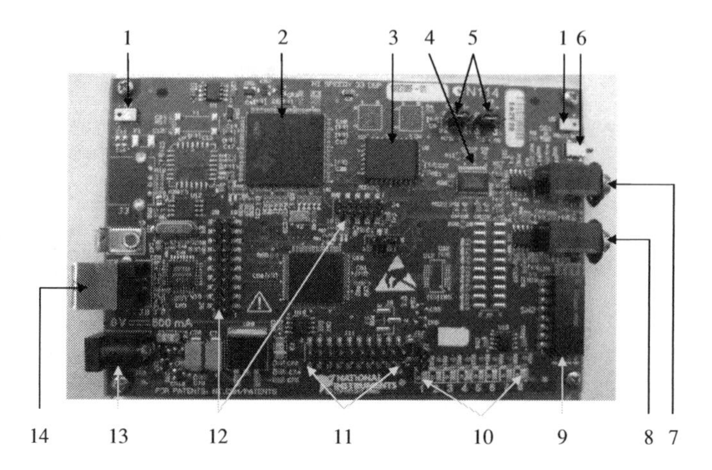

FIGURE 2.1 Top view layout of NI SPEEDY-33.

1. *Onboard Microphones*: There are two microphones available on the board which can be used for any audio input applications. The level of the input gain has to be set according to the microphone level setting. An audio input conditioning circuitry is connected to the onboard microphones but is mechanically disconnected if an external microphone is used.
2. *Digital Signal Processor (TMS320VC33)/On Chip Memory*

   DSP—The DSP present on the NI SPEEDY-33 is a flexible, easy-to-use and a powerful floating-point processor. It is a 32-bit processor capable of performing mathematical operations designed around CMOS technology having 0.18 μm. It supports 150 million floating point operations which can be executed per second.

   On Chip Memory—Flash and On Chip Memory are the two types of memory present on the NI SPEEDY-33 board. The On Chip Memory of 136 KB is used since it has a fast speed for algorithms. Flash Memory is explained below.
3. *Flash Memory*: This allows the device to be used in program and run mode without connecting it to a PC. It is a byte size and organized as 8 × 512 K mapping to a 32 × 512 K area.
4. *Stereo A/D, D/A*: The audio CODEC present onboard is 16 bits and dual channel having sampling of up to 48 KHz on the input signal. Sample rates of 8 KHz, 18 KHz, 24 KHz, 36 KHz, and 48 KHz are allowed by

the software components present on the DSP module for various applications.

5. *Audio Input Level Jumpers*: These consist of two channels, namely Left Channel AI Level Jumper and Right Channel AI Level Jumper. They control the gain of the input audio signals.
6. *Reset Button*: There is a small pushbutton on the front side of the board which serves as the reset button that resets the DSP.
7. *Audio Stereo Input Port/Output Port*: This can connect any audio input to its input port and speaker to its output port. External headsets can also be connected to the output port, but there is no control over the signal level or output gain.
8. *Switch Input Port*: There are eight switch inputs that are memory mapped and are used for general purpose inputs, which are accessed by DSP by reading a particular bit of the switch port.
9. *Digital Output Port LEDs*: There are eight LEDs that are memory mapped and are used for general purpose outputs.
10. *Simple Expansion Digital I/O Connector*: This consists of 20 pins which include ground, power, 8 digital inputs as well as outputs to easily interface external hardware. The pinout of the connector is shown in Figure 2.2. The inputs can tolerate 5 V, but the digital input/output signals are 3.3 V. The Reset pin (Pin 20) is driven low in order to reset the DSP.

5 V (Out)	1	2	3 V (Out)
IN1 (In)	3	4	OUT1 (Out)
IN2 (In)	5	6	OUT2 (Out)
IN3 (In)	7	8	OUT3 (Out)
IN4 (In)	9	10	OUT4 (Out)
IN5 (In)	11	12	OUT5 (Out)
IN6 (In)	13	14	OUT6 (Out)
IN7 (In)	15	16	OUT7 (Out)
IN8 (In)	17	18	OUT8 (Out)
GND	19	20	ResetLow (In)

FIGURE 2.2 Simple Expansion Digital I/O Connector (J11).

11. *Power Port*: This is used when the device is operated in standalone mode and not connected to a PC through the USB port. An input voltage with inner contact voltage as +ve VDC, 500 mA and outside contact as ground.

12. *PC USB Port*: This connects the host (PC) to the target (NI SPEEDY board) with a USB cable which makes the board operate at full speed.

Board setup

Step 1: Open the software National Instruments LabVIEW 8.6.

Step 2: Click on Launch LabVIEW as shown in Figure 2.3.

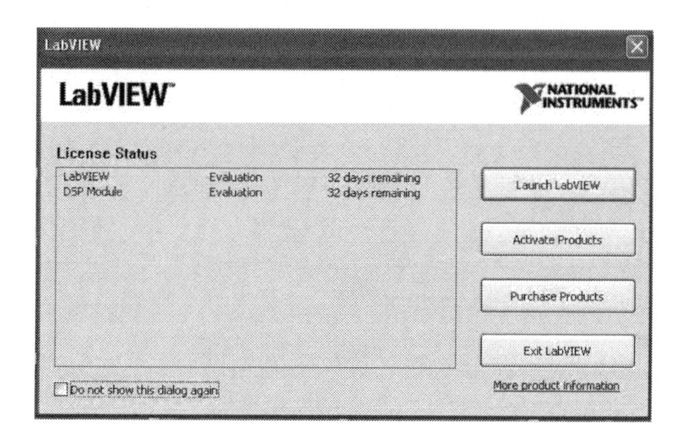

FIGURE 2.3 Launch LabVIEW.

Step 3: The screen shown in Figure 2.4 will appear.

FIGURE 2.4 Screenshot of LabVIEW setup.

Step 4: Select Targets as DSP Project from the drop down menu as shown in Figure 2.5 and click on Go.

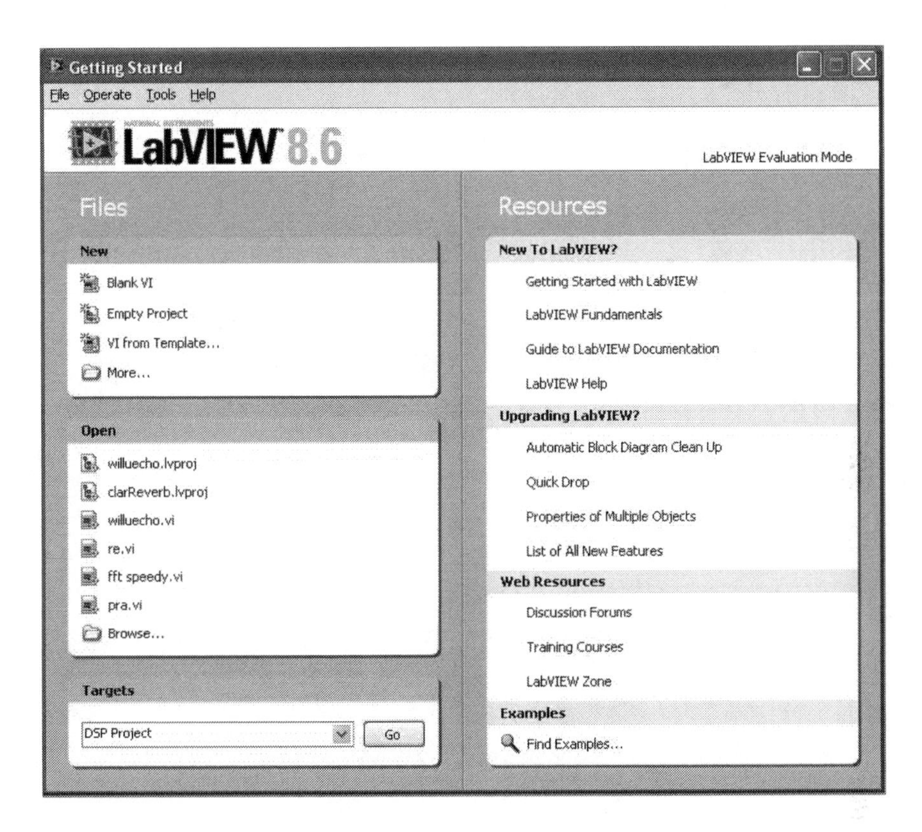

FIGURE 2.5 Getting Started.

Step 5: Select Project Type as "New DSP project, blank VI" as shown in Figure 2.6 and click on Next.

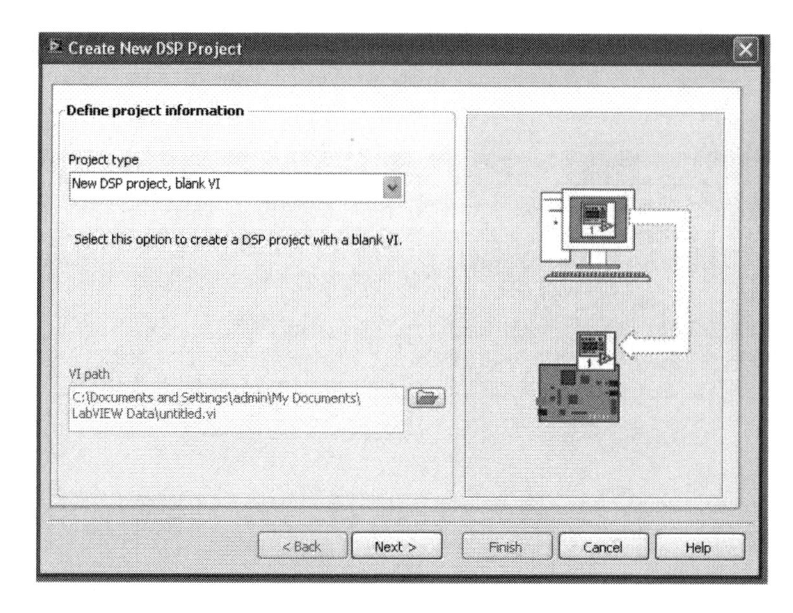

FIGURE 2.6 Create New DSP Project.

Step 6: Select the Target Type "SPEEDY-33" from the drop down menu as shown in Figure 2.7. Check all the options and click on Finish.

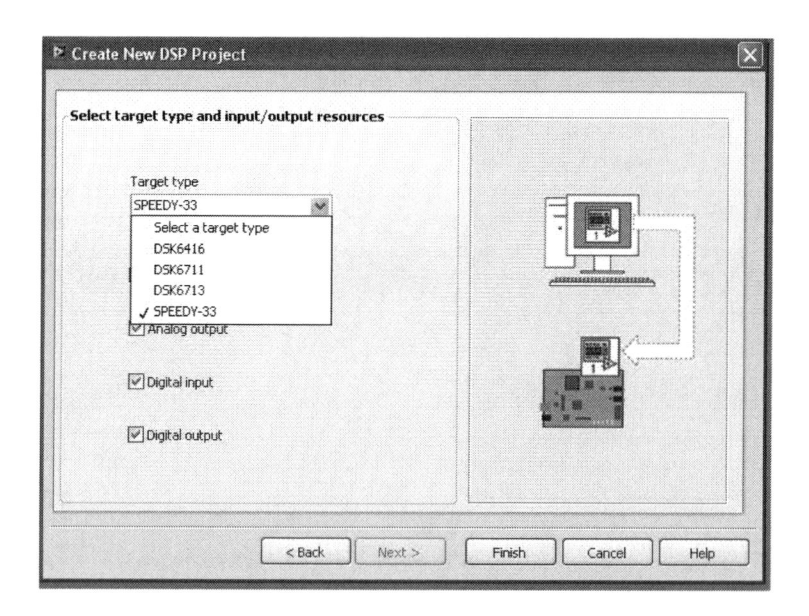

FIGURE 2.7 Select the target.

Step 7: Three windows will open, Block Diagram, Front Panel and a Project Explorer as shown in Figure 2.8.

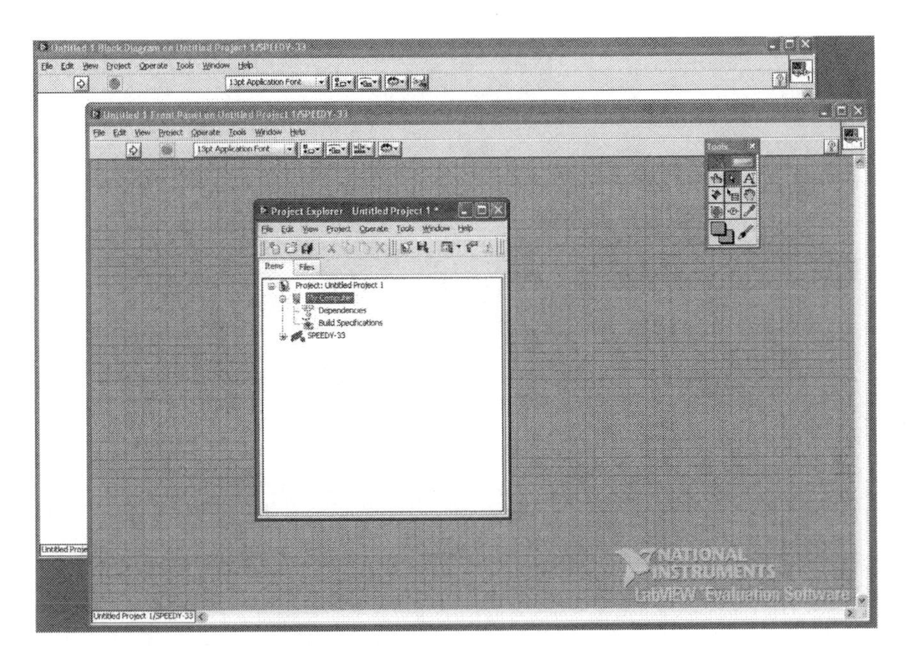

FIGURE 2.8 Block Diagram, Front Panel and Project Explorer.

Step 8: Go to the Block Diagram Window. Go to View and then select Functions palette. The Function palette will be seen on the side of the Block Diagram as shown in Figure 2.9.

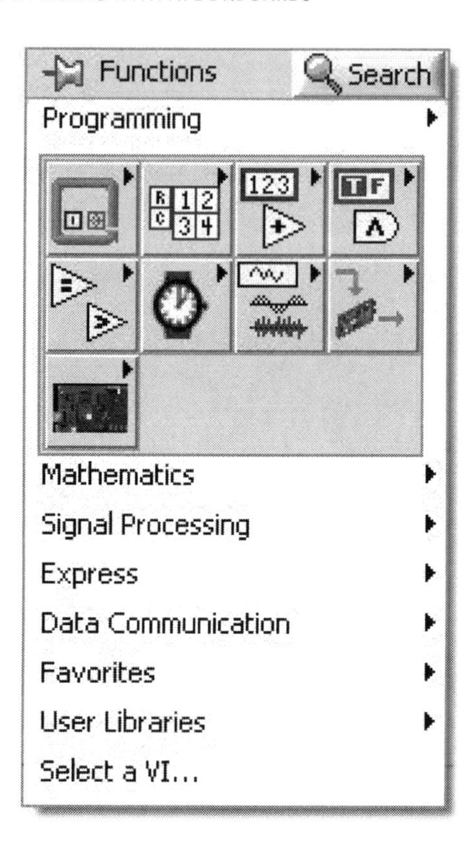

FIGURE 2.9 Functions Palette.

Step 9: Go to the Front Panel Window. Go to View and then select Controls palette. The Controls palette will be seen on the side of the Front Panel as shown in Figure 2.10.

FIGURE 2.10 Controls Palette.

Step 10: Connect the USB cable which is also the power cable between the NI SPEEDY-33 and computer.

2.2 NI ELVIS features

Features

The board we used is NI ELVIS II$^+$ having features such as 1 channel, 5 MHz function generator, 2 channel, 100 MS/s oscilloscope, 16 channel, 16 bit AI/AO, 24 digital IO lines [2]. The top view of NI ELVIS II$^+$ is shown in Figure 2.11.

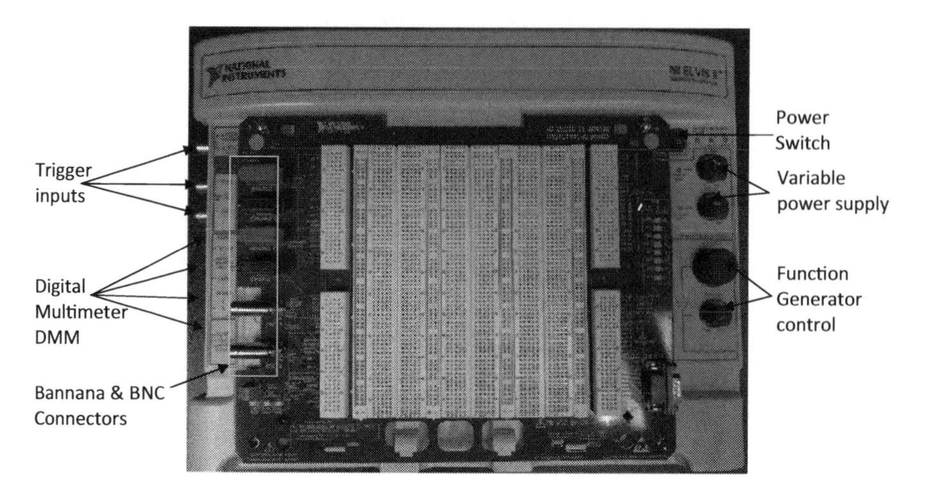

FIGURE 2.11 Top view of NI ELVIS II+.

Various Board Features:

1. *Digital Trigger Input/Function Generator Output and Oscilloscope Connectors*

   FGEN/Trigger Connector: It is an output of the function generator or an input to the digital trigger.

   Oscilloscope Connectors

   BNC Connector (CH0 & CH1): This is an input for channel 0 & 1.

2. *Digital Multimeter (DMM) Connectors*

   Red Banana Jack (Current): This is a positive input used for DMM current measurements.

   Black Banana Jack (Common): This is a common reference.

   Red Banana Jack (Diode, Resistance and Voltage): It is a positive input used for DMM in diode, resistance and voltage measurements.

3. *Function Generator—Manual Controls*: The amplitude and frequency of the function generator can be adjusted by manually turning the knobs.

4. *Variable Power Supply—Manual Controls*: Here the voltage of two power supplies can be adjusted. Supply+ provides voltage between 0 V and +12 V and supply- provides voltage between –12 V and 0 V. The knobs are active if the power supply is in manual mode and the LEDs glow near each knob as well.

5. *Power Switch of the Prototyping Board*: When the switch is ON, a power LED glows indicating power is provided to the prototyping board. If it is connected to the host, then the Ready switch should glow.
6. *NI ELVIS II Series Prototyping Board*: It is an area provided to build circuitry and has the required connections to connect signals for general applications.

Prototyping board features The breadboard and its peripherals are shown in Figure 2.12.

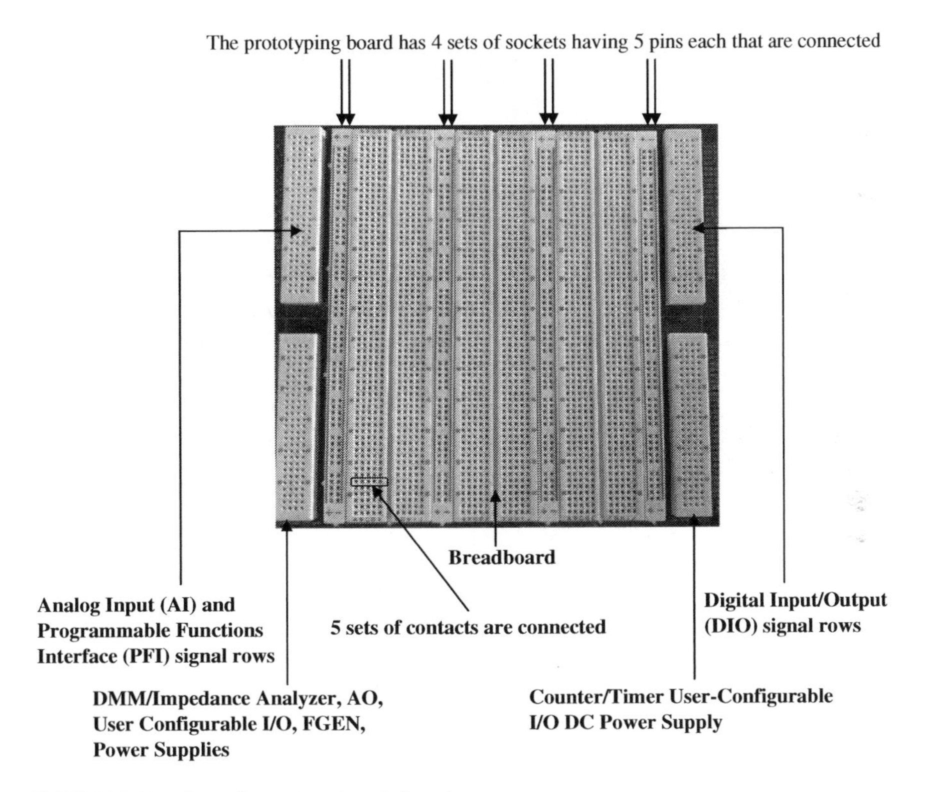

FIGURE 2.12 Breadboard and peripherals.

Breadboard: This is the area wherein circuits are built. Buses which are placed horizontally are labeled "+" used for power and "–" used for ground signals. The actual circuit is built on the vertical contacts.

1. *Analog Input (AI) and Programmable Functions Interface (PFI) signal rows*
   a. *Analog Inputs*: There are 8AI from AI0 to AI7. The "+" AI is connected to the positive pin socket and "–" AI is connected to the negative pin socket.
   b. *Programmable Functions (PF) I/O*: The PFI lines are labeled 0, 1, 2, 5, 6, 7, 10 and 11 and are used for AI, AO, timer/counter engines and can be used for static digital I/O as well.
   c. *AI GND and AI SENSE (INPUT)*: The pin sockets are used when the signal measured has a ground potential different from that of the workstation.
2. *DMM/Impedance Analyzer, AO, User Configurable I/O, FGEN, Power Supplies*
   a. *Digital Multimeter (DMM)/Impedance Analyzer*

      DUT+: This is a two wire excitation for impedance analyzer, inductance and capacitance measurement and a three wire excitation collector terminal current/voltage analyzer of a BJT.

      DUT–: This is a two wire excitation for impedance analyzer, inductance and capacitance measurement and a three wire excitation emitter terminal current/voltage analyzer of a BJT.

      BASE: This is a three wire excitation base terminal current/voltage analyzer of a BJT.
   b. *Analog Output (AO)*: There are two AOs, AO0 and AO1, used for the waveform generator as an output. A0 is used for three wire excitation base terminal current/voltage analyzer of a BJT.
   c. *User Configurable Input/Output*: There are four Banana jacks labeled BananaA, BananaB, BananaC and BananaD. There are two BNC connectors, BNC1 and BNC2. There are eight LEDs which are bicolor wherein the anode is connected through a 220 Ω resistor to a distribution strip and the cathode is grounded.
   d. *Digital Input/Output (DIO) signal rows*: There are 24 DIOs labeled from DIO0 to DIO23 which can be used to read or write data digitally (0 V–5 V). These are connected to port 0 and can be programmed to be used as inputs or outputs by means of Soft Front Panels (SFPs).

e. *Counter/Timer User-Configurable Input/Output*: There are two counter/timer available on the prototyping board which can also be accessed by software. These are used as inputs for pulse generation, edge detection and counting TTL signals.

Board setup

Step 1: Use the USB cable to connect the NI ELVIS workstation to the computer.

Step 2: Connect the AC/DC power supply to the NI ELVIS II+ Workstation. Then, plug the power supply to a socket and press the power switch button on the back of the workstation to ON position.

Step 3: Position the opening in the prototyping board over the prototyping board mounting bracket.

Step 4: Slide the edge connector of the prototyping board into the receptacle on the workstation.

Step 5: Gently rock the board to ease it into place. It may be a tight fit, but do not force the board into place. Slide the prototyping board into the prototyping board mounting bracket.

Step 6: Now click on the NI Launcher icon on the Desktop PC to open the NI ELVIS Instruments launcher.

Step 7: The different integrated instruments available on NI ELVIS II+ are shown in Figure 2.13.

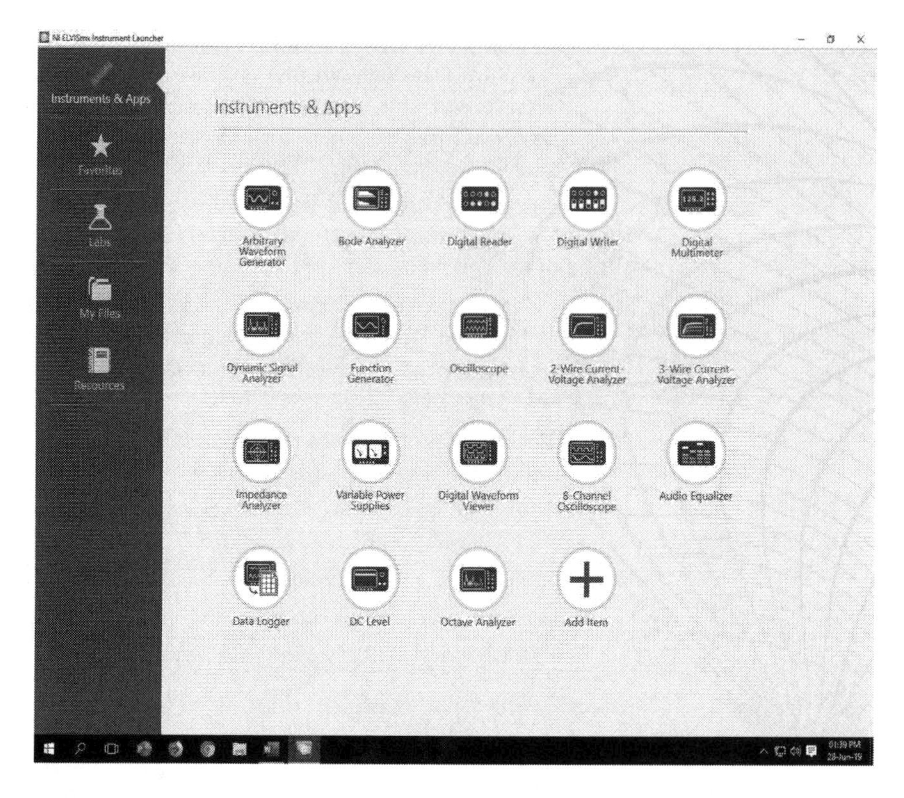

FIGURE 2.13 NI ELVISmx Instrument Launcher.

2.3 myRIO

Features

NI myRIO-1900 is a convenient Reconfigurable Input Output (RIO) device that is used to design and control mechatronics and robotics systems. NI myRIO-1900 offers Digital Input & Output (DIO), Analog Input (AI), Analog Output (AO), power and audio output in an embedded device [3]. NI myRIO-1900 is connected to a host computer through a USB cable or wireless 802.11 b, g, n. and is shown in Figure 2.14.

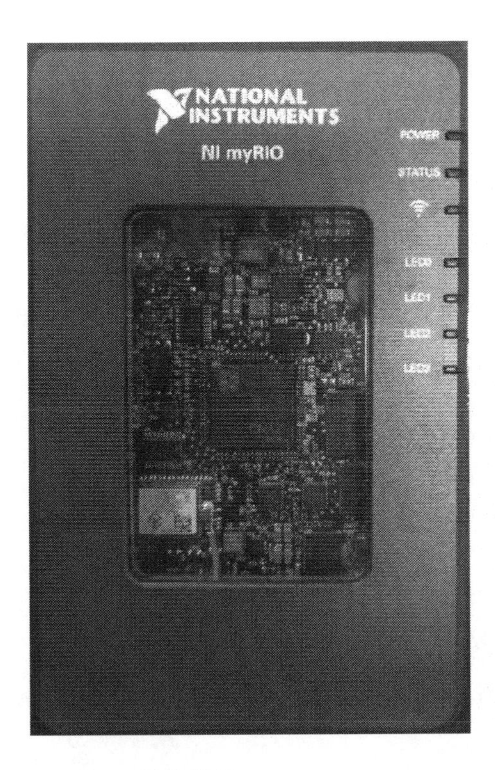

FIGURE 2.14 NI myRIO-1900.

The functions and arrangement of the components are shown in Figure 2.15.

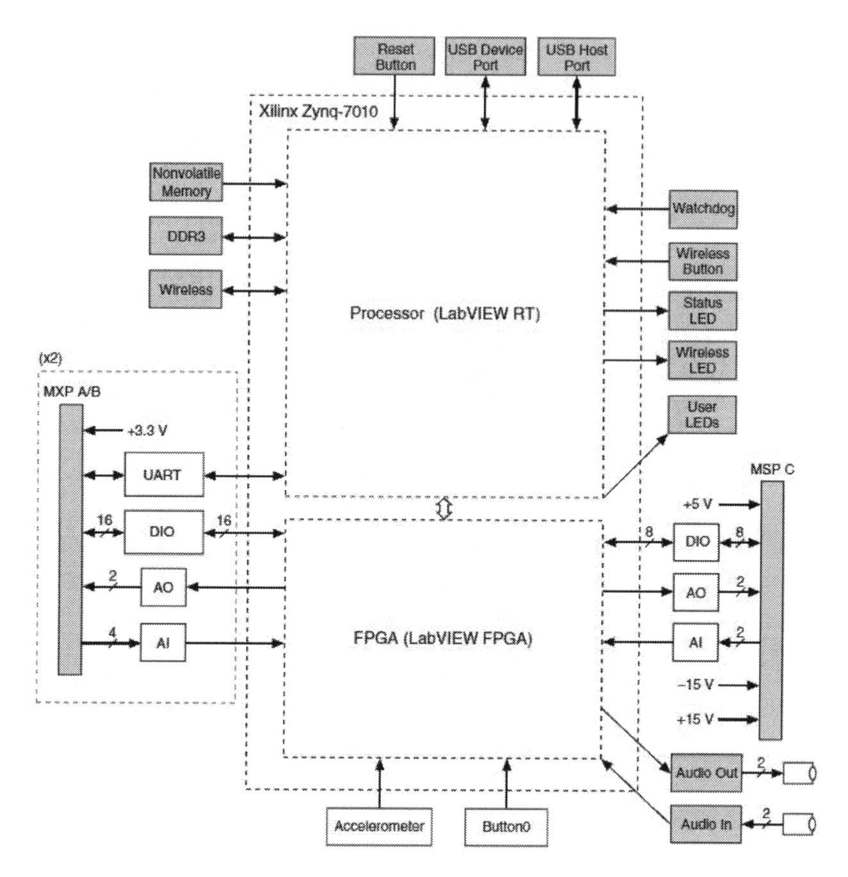

FIGURE 2.15 Block Diagram of NI myRIO-1900 Hardware.

myRIO Expansion Port (MXP) Connector: The MXP connectors A and B transmit a similar set of signals. In software, the signals are known as ConnectorA/DIO1 and ConnectorB/DIO1 for connectors A and B respectively. The signals of the MXP connectors of A and B are shown in Figure 2.16. Some pins have a secondary function besides the main function.

FIGURE 2.16 Main/secondary signals of MXP Connectors A and B (left side view).

It has a power output of +5 V with AGND as reference and a power output of +3.3 V with DGND as reference. AI 0, AI1, AI2 and AI3 are analog inputs which are single ended channels and have a voltage range between 0 and 5 V. AO0 and AO1 are analog outputs which are single ended channels and have a voltage range between 0 and 5 V. DIO0 up to DIO15 are the digital input output with a 3.3 V for output and a 3.3 V/5 V for input. UART.TX is the UART transmit output whereas UART.RX is the UART receive input. The UART lines are electrically similar to the DIO lines. AGND is the reference for AI and AO signals whereas DGND is the reference for DIO signals.

Mini System Port (MSP) Connector: MSP connector C is shown in Figure 2.17. Some pins have a secondary function besides the main function.

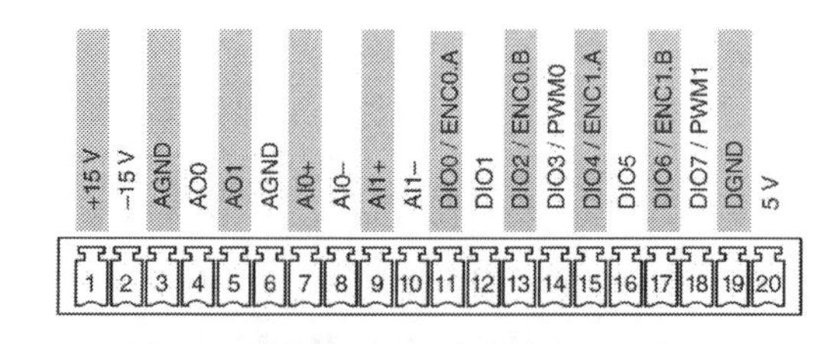

FIGURE 2.17 Main/secondary signals on MSP Connector C (right side view).

It has a power output of +15 V/–15 V with AGND as reference and a power output of +5 V with DGND as reference. AGND is the reference for AI and AO signals, whereas DGND is the reference for DIO signals. AI0+, AI1+, AI0– and AI1– are the differential analog inputs which have a voltage range between –10 V and +10 V. AO0 and AO1 are analog outputs which are single ended channels and have a voltage range between –10 V and +10 V. DIO0 up to DIO7 are the digital input output with a 3.3 V for output and a 3.3 V/5 V for input. Audio In is the left and right AI on stereo connector whereas Audio Out is the left and right AO on stereo connector.

Analog Input Channels: NI myRIO-1900 has AI channels on the stereo AI connector, MXP connectors A and B and MSP connector C. The AI is multiplexed to a single ADC which samples all the channels. The MXP connectors A and B have four single-ended AI channels (AI0–AI3) per connector that can be used to measure 0–5 V signals. The MSP connector C has two high-impedance differential AI channels (AI0 and AI1) that can be used to measure signals up to ±10 V. The AI are right and left stereo line-level inputs with a range from –2.5 V to +2.5 V.

Analog Output Channels: NI myRIO-1900 has AO channels on the stereo AO connector, MXP connectors A and B and MSP connector C. The AO channel has a DAC which updates all the channels. The DACs of the AO channels are managed by two serial communication buses of the FPGA. MXP connectors A and B are both allocated a single bus whereas the AO and MSP connector C are allocated another bus. The MXP connectors A and B have 2 AO channels (AO0 and AO1) per connector that can be used to create 0–5 V signals. The MSP connector C has 2 AO channels (AO0 and AO1) that can be used to create signals up to ± 10 V. The AO are right and left stereo line-level outputs which are used to drive headphones.

Accelerometer: NI myRIO-1900 consists of a 3-axis accelerometer which samples every axis constantly and updates the result in the register.

DIO Lines: NI myRIO-1900 MSP and MXP connectors have general-purpose DIO lines with 3.3 V. Sixteen DIO lines are available on MXP connectors A and B each. DIO0 to DIO13 have a 40 kΩ pullup resistor, 3.3 V, whereas DIO14 and DIO15 have a 2.1 kΩ pullup resistor, 3.3 V. Eight DIO lines are available on the MSP connector C of which all the DIO lines have a 40 kΩ pulldown resistor to GND. The reference of the DIO lines is DGND, which can be programmed as inputs or outputs. The secondary functions of the connectors are quadrature encoder input, pulse-width modulation (PWM), I2C and Serial Peripheral Interface Bus (SPI). If a DIO line is left floating, then it floats in the same direction as that of a pull resistor. It can float due to the following situations: If the line is constituted as an input or if the myRIO board is starting up or powering down. A stronger resistor can be added to float the DIO line in the opposite direction.

UART Lines: NI myRIO-1900 has a UART transmitting output line and a UART receiving input line on every MXP connector. The UART lines are electrically similar to DIO lines DIO0 to DIO13 of the MXP connectors. Similarly UART.TX and UART.RX have 40 kΩ pullup resistors up to 3.3 V. LabVIEW Real-Time is used to write and read over the UART lines.

Board setup

Step 1: Open the software National Instruments LabVIEW myRIO 2016.

Step 2: Click on Launch LabVIEW

Step 3: The screen LabVIEW myRIO 2016 will appear.

Step 4: Click on myRIO Project under Create Project as shown in Figure 2.18.

FIGURE 2.18 Creating a project.

Step 5: Now the target board attached to PC is searched automatically.

Step 6: Specify the project name and project root once the myRIO is found as shown in Figure 2.19.

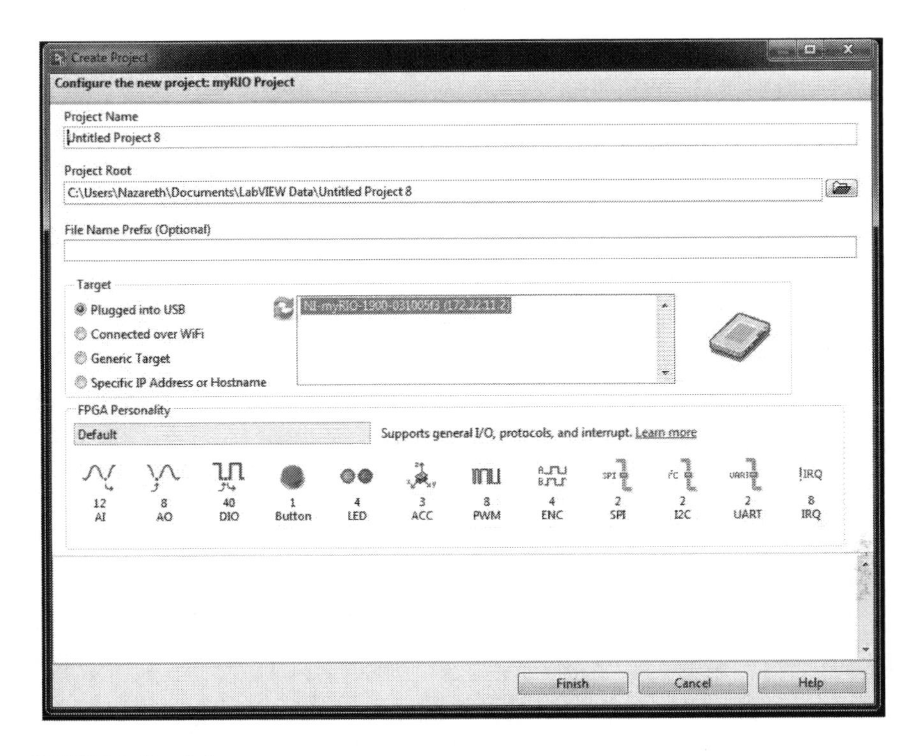

FIGURE 2.19 Select target.

Step 7: A Project Explorer will appear then Click on the "+" sign such that Main.vi is visible.

Step 8: The figure shown in Figure 2.20 will appear in which the Project Explorer, Block Diagram and Front Panel will be seen.

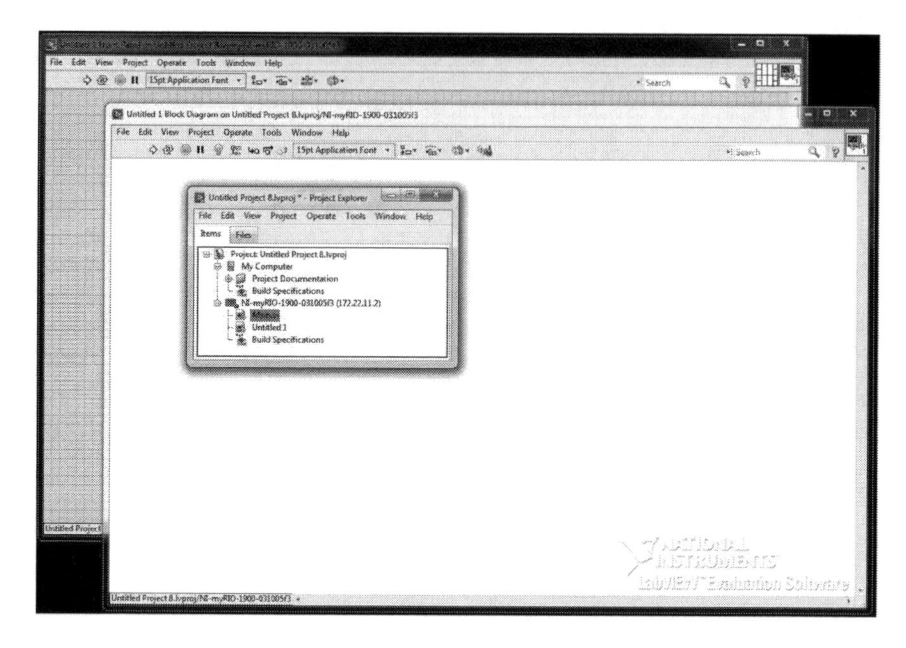

FIGURE 2.20 Front Panel, Block Diagram and Project Explorer.

Step 9: Right click on the Block Diagram such that the Functions palette will appear then right click on the Front Panel such that the Controls palette will appear.

References

1. L.J. Karam and N. Mounsef. *Introduction to Engineering: A Starter's Guide with Hands-On Digital Multimedia and Robotics Explorations.* Morgan & Claypool Publishers: San Rafael, CA, 2008.
2. *NI ELVIS II Series User Manual.* National Instruments Corporation.
3. *User Guide and Specifications*, NI myRIO-1900, May 2016.

CHAPTER THREE

Exploring SPEEDY-33

3.1 LEDs and Switches

In this section, the digital outputs connected to LEDs and the digital inputs to switches are interfaced using the SPEEDY-33. The board consists of eight LEDs and eight DIP switches, which can be configured using the DSP module. Writing a "1" to the LED port enables the LED to glow. Similarly the switch can be enabled by writing a "1" to the switch port.

Implementation Steps

Step 1: In the Block Diagram window, right click on the screen. Under the Functions dialog box, click on Programming and then select Elemental I/O. Now click on DSP Switch.vi as shown in Figure 3.1.

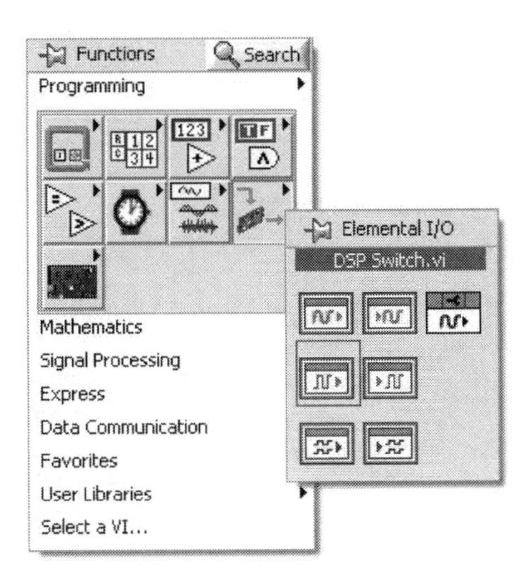

FIGURE 3.1 DSP Switch.vi.

Step 2: Right click on the DSP Switch.vi and select Create and then click on Indicator.

Step 3: Right click on the Block Diagram and then select Functions. Now select Programming, Structures and click on While Loop as shown in Figure 3.2. Now click and drag from the left top corner to the bottom right corner to form a rectangle.

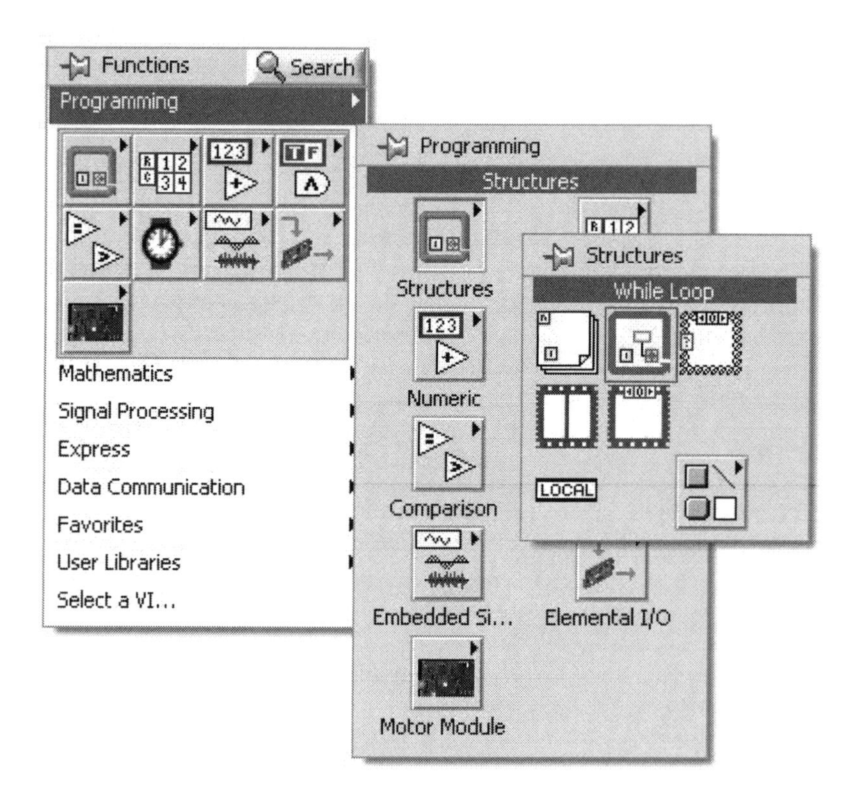

FIGURE 3.2 While Loop.

Step 4: Place a while loop around the switch and run the program. Only a single switch can be toggled. If Switch 1 is in the ON position, Switch 1 will glow in the Front Panel; if the Switch is in the OFF position, Switch 1 will turn off.

Step 5: In the Block Diagram window, right click on the screen. Under the Functions dialog box, click on Programming and then select Elemental I/O. Now click on DSP LED.vi as shown in Figure 3.3.

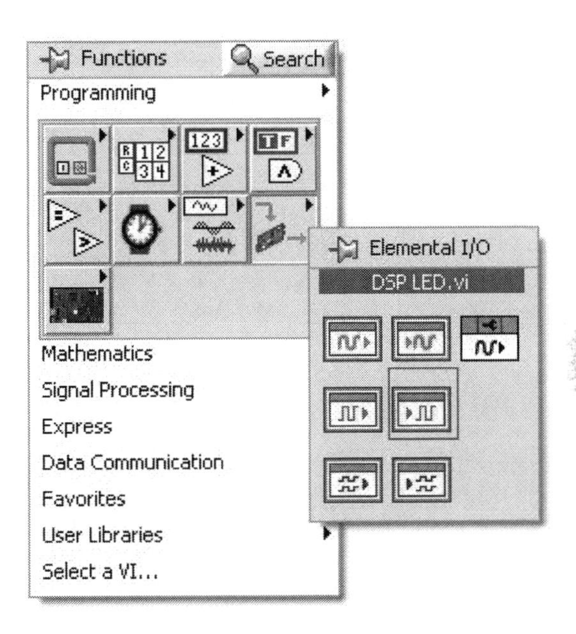

FIGURE 3.3 DSP LED.vi.

Step 6: Right click on the DSP LED.vi and select Create and then Control.

Step 7: The block diagram with Single Switch and LED with indicator and control is created. Run the program, press the first switch in one direction on the board and notice the LED turn ON and turn OFF in the other direction. If LED 1 in the Front Panel is put ON, the first LED on the board will turn ON and vice versa.

Step 8: Now delete the Control attached to the LED and connect the LED directly to the Switch as shown in Figure 3.4.

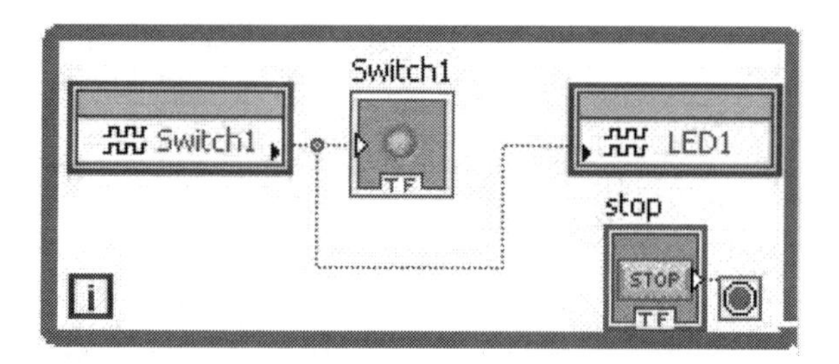

FIGURE 3.4 Block Diagram with Single Switch and LED with only indicator.

Step 9: Add upto eight switches and eight LEDs and connect them as shown in Figure 3.5 by selecting DSP Switch Bank and DSP LED Bank under Elemental I/O or by dragging the bottom edge downward until eight switches and eight LEDs are visible. Now all the switches present on the board can be used and displayed on the LEDs.

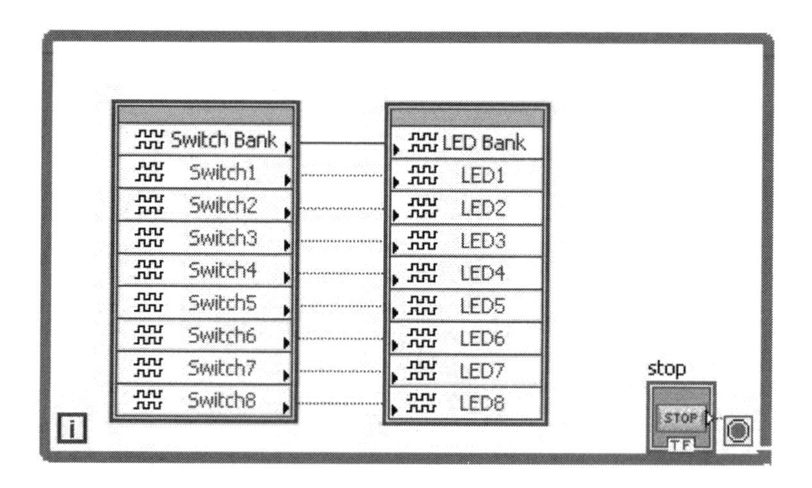

FIGURE 3.5 Block Diagram with eight switches and eight LEDs.

3.2 Keypad Interfacing

In this section, a DTMF signal is generated which is a sum of two pure sine signals that can be used in the keypad of a telephone. Here none of the frequencies of any two keys are multiples of each other. The frequencies are chosen such that they lie in the voice communication range of 300–3400 Hz. Seven different frequencies are used to generate 12 digits, namely 0–9, * and # that are found on a keypad of a telephone [1].

Implementation Steps

Step 1: In the Block Diagram window, right click on the screen. Under the Functions dialog box, click on Programming and then select Embedded Signal Generation. Now click on Simulate Signal as shown in Figure 3.6.

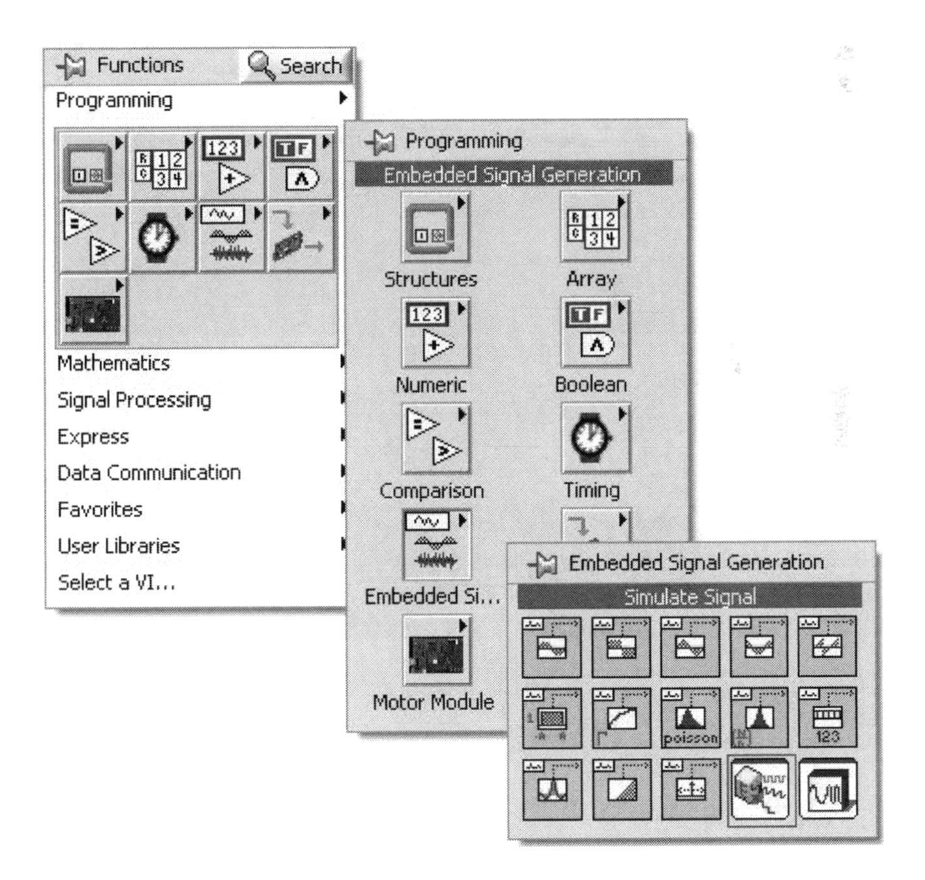

FIGURE 3.6 Simulate Signal.

Step 2: In the Block Diagram window, right click on the screen. Under the Functions dialog box, click on Programming and select Numeric. Now click on Add.

Step 3: Right click on the Block Diagram and then select Functions. Now select Programming, Elemental I/O and click on Analog Output.vi.

Step 4: Right click on the Block Diagram and then select Functions. Now select Programming, Array and click on Index Array.

Step 5: Right click on Index Array at the array input, select Create and click on Constant. After creating constants for each of the index arrays, drag the indices of the index array such that 13 indices are visible. Initialize the indices, as shown in Figure 3.7, which correspond to the output frequencies of each keypad digit.

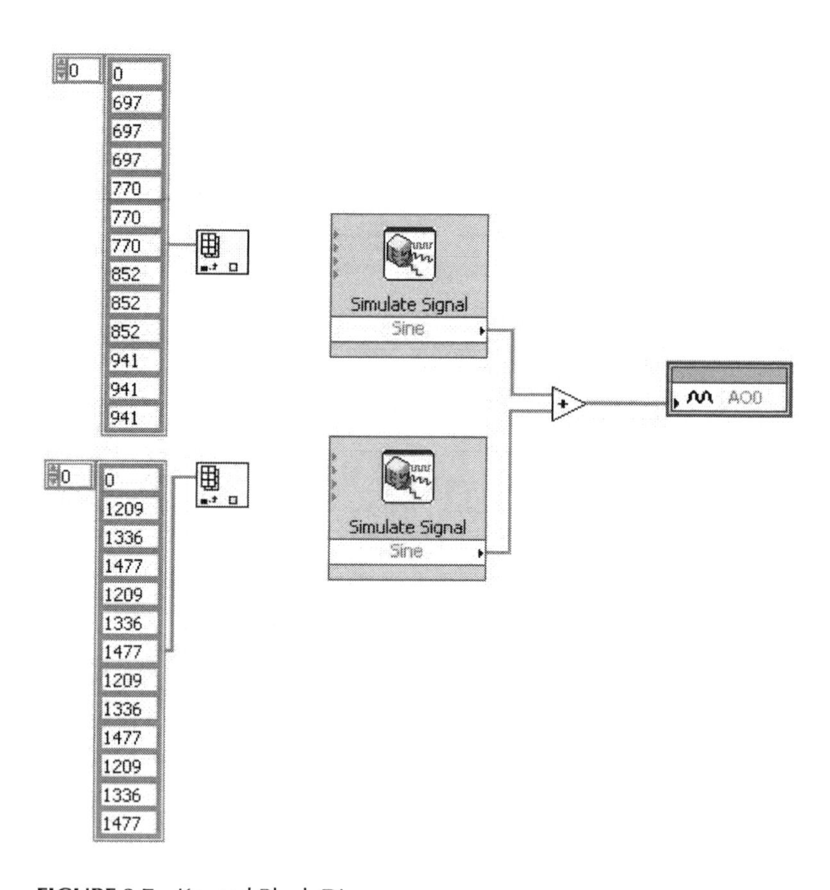

FIGURE 3.7 Keypad Block Diagram.

Row 1 has a frequency of 697 Hz, row 2 has 770 Hz, row 3 has 852 Hz and row 4 has 941 Hz, whereas column 1 has a frequency of 1209 Hz, column 2 has 1336 Hz and column 3 has 1477 Hz. In order to generate the output frequency for digit 1, the first row and first column frequency has to be combined as shown in Table 3.1. Therefore keypad digit 1 is 697 + 1209 = 1906 Hz. The other keypad frequencies are generated in the same manner. The frequencies are chosen such that none of the frequencies are multiples of each other.

Table 3.1 Combination of DTMF frequency

Sr. no.	Keypad digit	O/P frequency (Hz)	
1	1	697 + 1209	1906
2	2	697 + 1336	2033
3	3	697 + 1477	2174
4	4	770 + 1209	1979
5	5	770 + 1336	2106
6	6	770 + 1477	2247
7	7	852 + 1209	2061
8	8	852 + 1336	2188
9	9	852 + 1477	2329
10	0	941 + 1209	2150
11	*	941 + 1336	2277
12	#	941 + 1477	2418

Step 6: Right click on the front panel and then select Controls. Now select Modern, Boolean and click on the OK button. Place 12 OK buttons in the front panel.

Step 7: Right click on the OK button and select Properties then deselect the Visible OK Button under Label and Change the Off text from OK to 1.

Step 8: Continue changing the properties of the remaining 11 OK buttons from OK to 2–9, 0 as well as * and #, so that it looks like Figure 3.8.

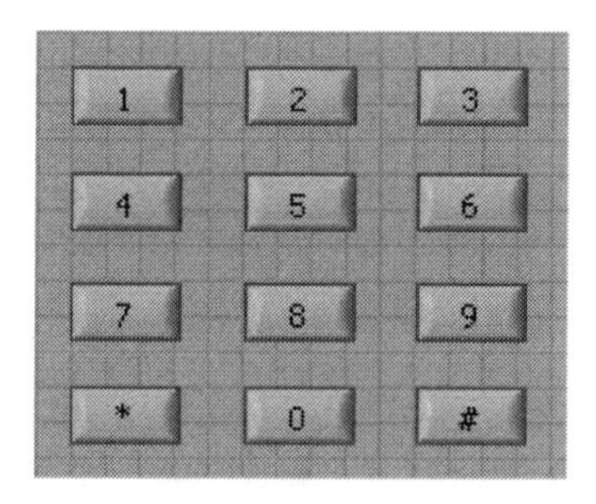

FIGURE 3.8 Keypad Front Panel.

Step 9: In the Block Diagram window, right click on the screen. Under the Functions dialog box, click on Programming and then select Comparison which has T/F.

Step 10: Add a Select comparison for every OK Button. Connect the Select blocks to the OK buttons as shown in Figure 3.9. Now right click on the T of the select and click on create constant and label it as 1 if the OK button is of digit 1. In the same manner, label it as 2 for OK button 2 and so on up to OK button 12. Further, right click on the F of the Select and click on create constant and leave the label as 0 for all the OK buttons.

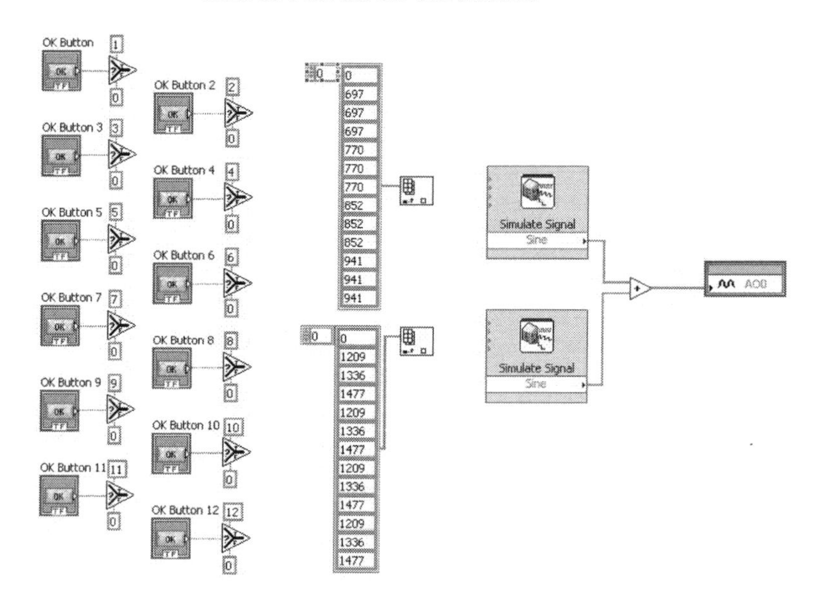

FIGURE 3.9 Comparison Block Diagram.

Step 11: Repeat Step 2 and add 11 more Add blocks to the block diagram and connect them as shown in Figure 3.10.

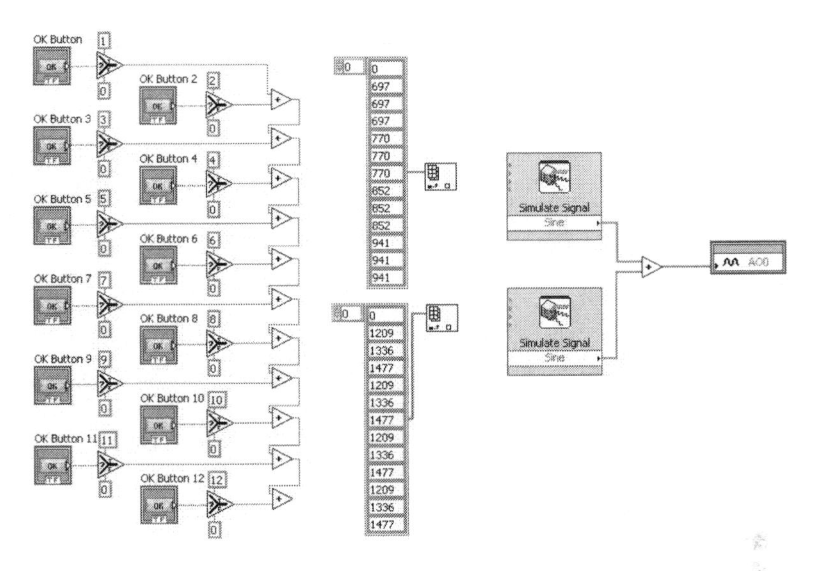

FIGURE 3.10 Add blocks in Block Diagram.

Step 12: In the Block Diagram window, right click on the screen. Under the Functions dialog box, click on Programming and then select Structures. Now click on For Loop.

Step 13: Add the For Loop around the Simulate Signal and Analog Output.vi as shown in Figure 3.11. Connect the input frequency of the Simulate Signal to the output of the Index Array.

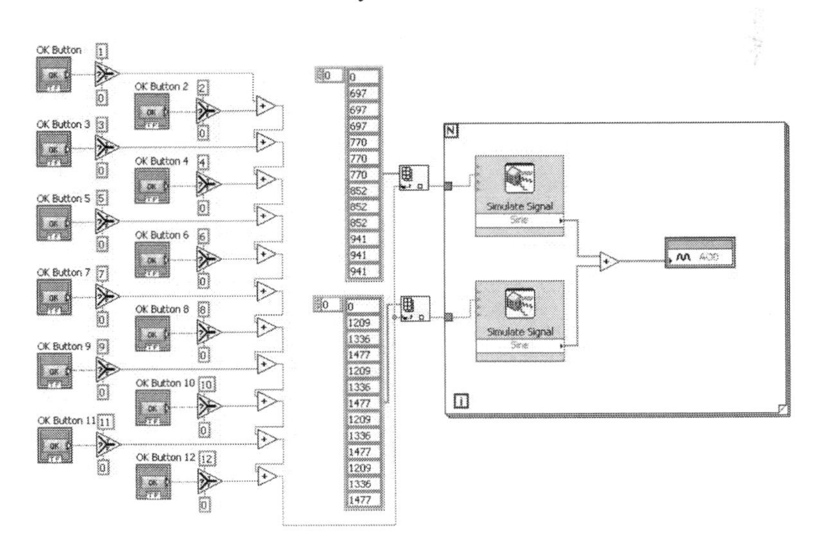

FIGURE 3.11 For Loop addition to Block Diagram.

Step 14: In the Block Diagram window, right click on the screen. Under the Functions dialog box, click on Programming and then select Numeric. Now click on Divide.

Step 15: Connect the Divide to the No. of Iterations in the For Loop as shown in Figure 3.12. Now right click on one of the inputs of the Divide and create constant with label equal to 150 since the minimum tone length while dialing a telephone number should be 40 ms and the other input as 16 to get the loop speed for one iteration since Loop Speed = (No. of Samples)/(Sampling Rate (Hz)) = 128/8000 = 16 ms.

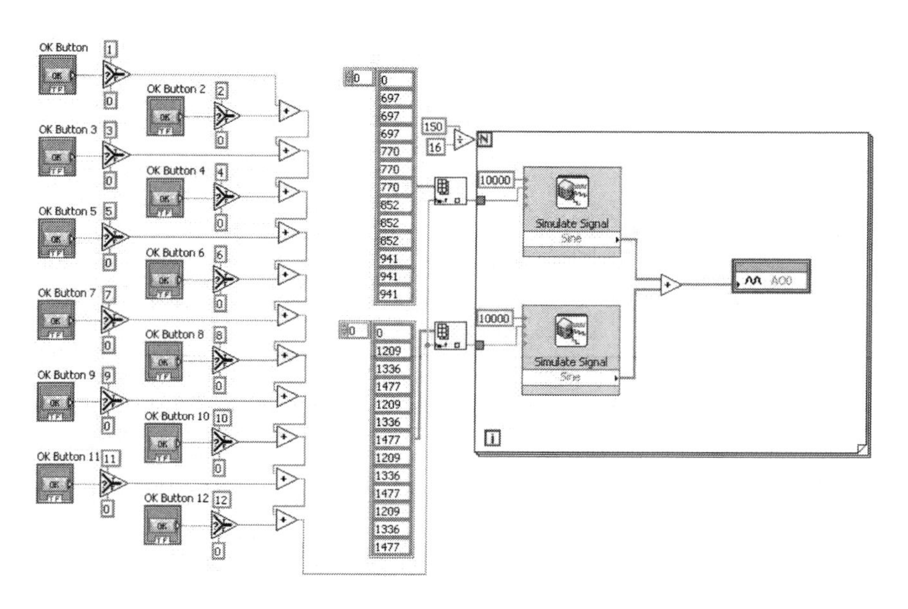

FIGURE 3.12 Divide in Block Diagram.

Step 16: In the Block Diagram window, right click on the screen. Under the Functions dialog box, click on Programming and then select Structures. Now click on While Loop.

Step 17: Drag the While Loop around the entire block as shown in Figure 3.13.

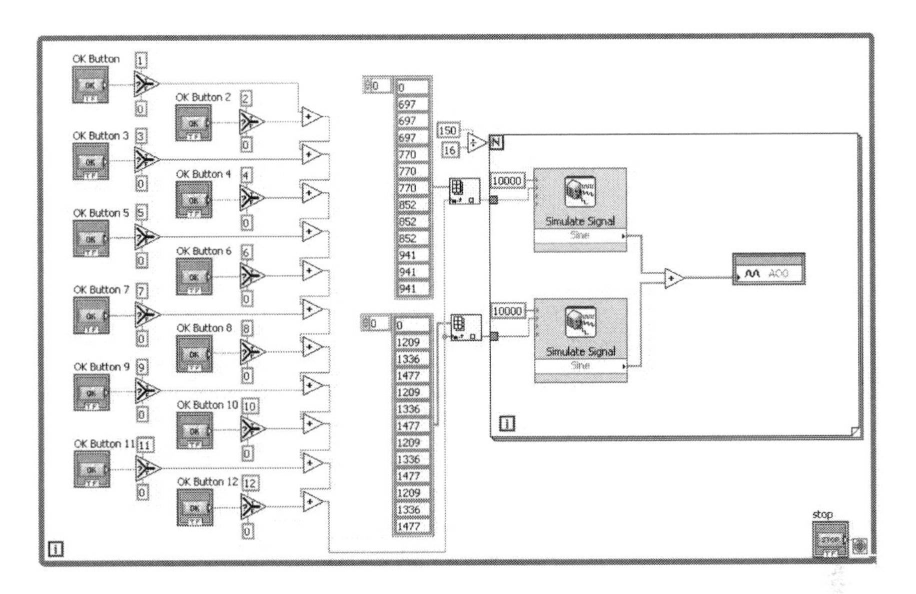

FIGURE 3.13 Final Block Diagram.

Step 18: The Front Panel will appear. Now Run the VI and try pressing the different keypad digits and note the sounds.

3.3 A/D and D/A conversion

In this section, the analog signals are converted to digital ones using an A/D converter and the digital signals are converted into analog ones using a D/A converter. If both the converters are interfaced to each other, the original signal can be obtained. The analog input can be obtained from the computer by playing a song on it, converted to digital output in the SPEEDY-33 and back to the user through earphones using the D/A converter [2].

Implementation Steps

Step 1: Right click on the Block Diagram and then select Functions. Now select Programming, Elemental I/O and click on Analog Input.vi. Right click on Analog Input.vi and select Properties; the Elemental I/O Properties dialog box will open. Under IO Node Configuration, select Sample Rate in Hz as 48000, Framesize as 256 and Gain as 1. Click on OK.

Step 2: Right click on the Block Diagram and then select Functions. Now select Programming, Elemental I/O and click on Analog Output.vi.

Step 3: Right click on the Block Diagram and then select Functions. Now select Programming, Structures and click on While Loop. Now click and drag from the left top corner to the bottom right corner to form a rectangle.

Step 4: Connect the blocks AI0 to AO0.

Step 5: In order to run the program, click on Run below the Menu bar. Connect the AUX cable from the PC to the Audio In of the board and the headphones to the Audio Out of the board. Speak on the mic present on the board and listen through the earphones or headphones.

Step 6: To view results on a graph, right click on the Front Panel and then select Controls. Now select Modern, Graph and click on Waveform Graph.

Step 7: After placing the waveform graph in the front panel, double click on the label Waveform Graph and change it to Input Signal.

Step 8: Now connect the Analog Input.vi to the Input Signal as shown in Figure 3.14.

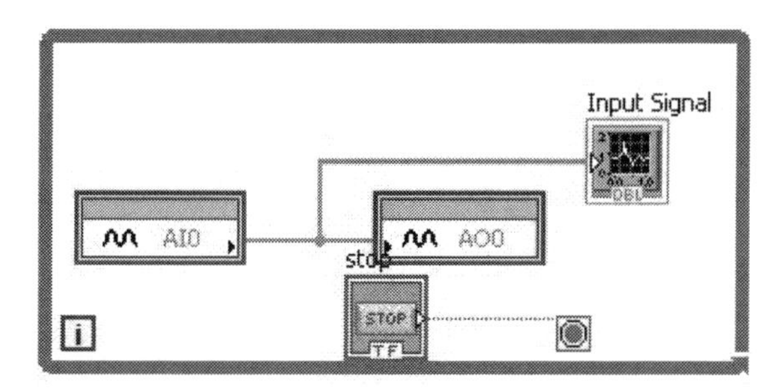

FIGURE 3.14 Final A/D and D/A Block Diagram.

Step 9: Now Run the VI.

3.4 Digital Filter Design

In this section, digital filters are designed in order to remove noise from a distorted signal using DSP Module and SPEEDY-33. Digital filters consist of digital components which use digital inputs to give digital outputs, unlike analog filters that consist only of analog components which use analog inputs to give analog outputs. There are various types of filters, namely lowpass, highpass, bandpass, bandstop, etc. A lowpass filter passes only low frequencies whereas a highpass filter passes only high frequencies. A bandpass filter passes a band of frequencies, and a bandstop filter rejects a band of frequencies.

Implementation Steps

Step 1: In the Block Diagram window, right click on the screen. Under the Functions dialog box, click on Programming and then select Embedded Signal Generation. Now click on EMB Sine Waveform.vi.

Step 2: In the Front Panel window, right click on the screen. Under the Controls dialog box, click on Modern and then select Numeric. Now click on Knob.

Step 3: Repeat Step 2 and place another knob in the front panel.

Step 4: In the Front Panel, right click on the screen and then select Controls. Now select Modern, Graph and click on Waveform Graph.

Step 5: Rename one of the Knobs as Frequency and the other as Amplitude.

Step 6: After connecting the blocks, right click on the Block Diagram and then select Functions. Now select Programming, Structures and click on While Loop.

Step 7: Place a while loop by clicking and dragging from the left top corner to the bottom right corner to form a rectangle around the components as shown in Figure 3.15.

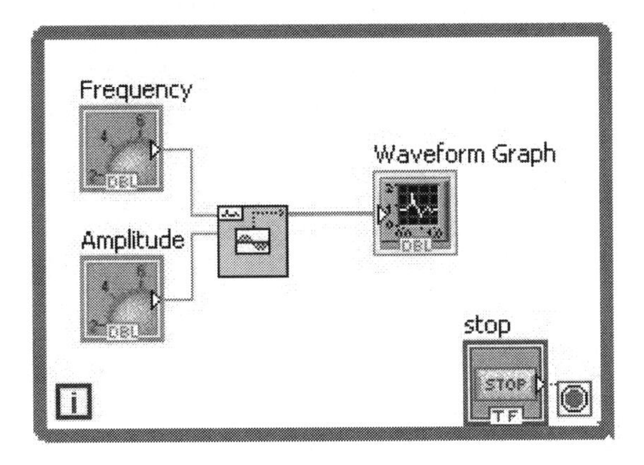

FIGURE 3.15 Filter design Block Diagram.

Step 8: Right click on the Block Diagram and then select Functions. Now select Signal Processing, Filters and click on Filter.

Step 9: Right click on the Filter and select Properties. A configure Filter dialog box will appear. Select Lowpass filtering type with a sampling rate of 8000, Cutoff Frequency 500 Hz, IIR filter with third order Butterworth filter. Click on OK.

Step 10: Right click on the Frequency Knob and Select Properties. Under the Scale tab, change the maximum scale range to 1000 and click on OK.

Step 11: Now connect the block as shown in Figure 3.16.

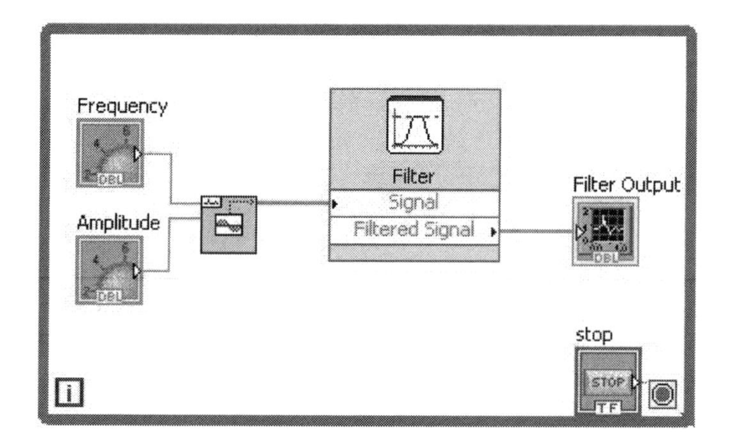

FIGURE 3.16 Block Diagram of a Lowpass Filter.

Step 12: Only low frequencies are passed as shown in Figure 3.17 and higher frequencies are not passed as shown in Figure 3.18 where the amplitude reduces to 0.6 p-p.

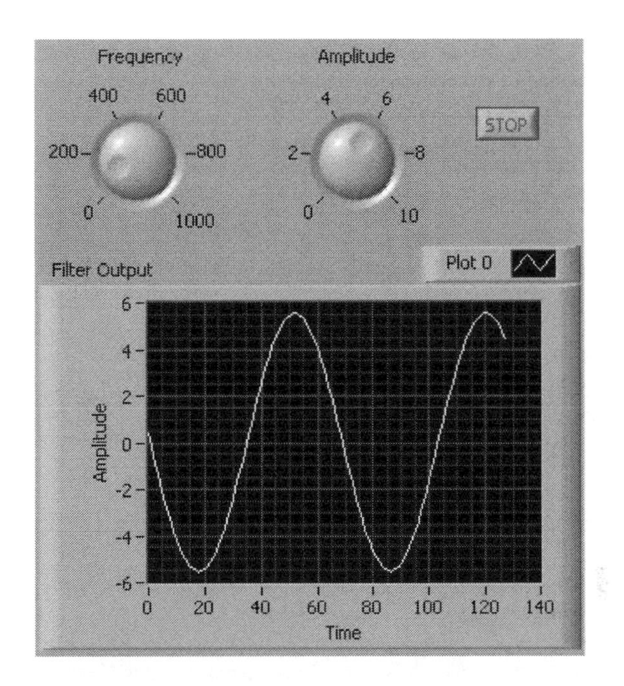

FIGURE 3.17 Front Panel, Lowpass Filter passing low frequencies.

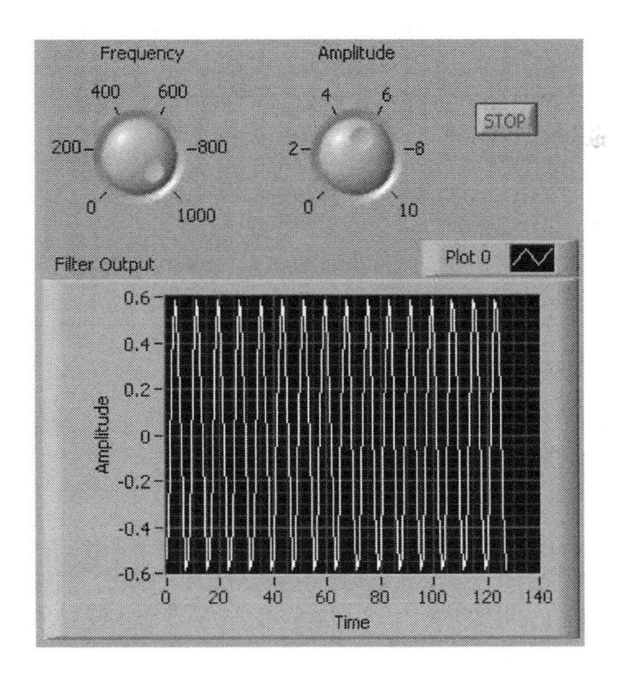

FIGURE 3.18 Front Panel, Lowpass Filter rejecting high frequencies.

Step 13: For a high pass filter, right click on the Filter and click on Properties. Change the filtering type to Highpass and Cutoff Frequency (Hz) to 1000. Click on OK. Change the Scale of the Frequency Knob to a maximum of 5000 and run the VI. Only high frequencies are passed.

Step 14: In order to make a bandpass or a bandstop Filter, place a lowpass and a highpass filter in series to form a bandpass or a bandstop filter as shown in Figure 3.19.

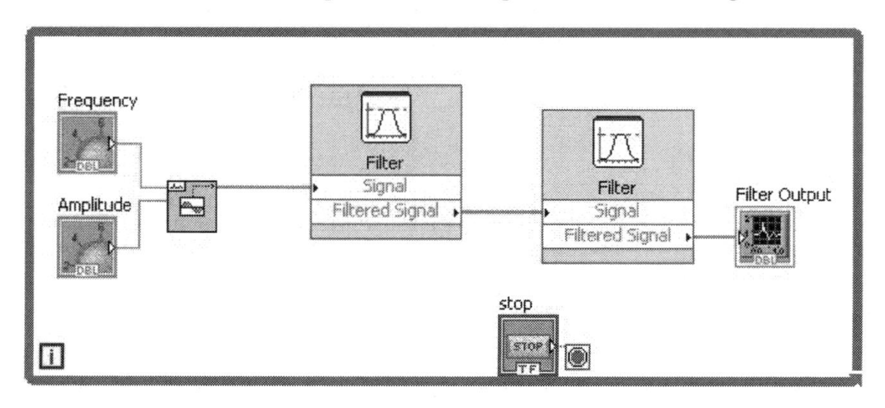

FIGURE 3.19 Block Diagram of a Bandpass Filter.

3.5 AM Modulation and Demodulation

An Analog communication system needs to transmit data using signal modulation to ensure the authenticity of the signal. There are various types of modulation, namely, Amplitude Modulation (AM), Frequency Modulation (FM) and Phase Modulation (PM). In this section, AM modulation is used wherein the amplitude of a carrier signal is modulated before sending it over a channel. AM demodulation retrieves the original signal by using a product detector wherein the carrier is multiplied with the AM signal and then passing it through a lowpass filter.

Implementation Steps

Step 1: In the Block Diagram window, right click on the screen. Under the Functions dialog box, click on Express and then select Input. Now click on Frequency Sweep Generator.

Step 2: Right click on Frequency Sweep Generator and click on Properties.

Step 3: Change the Amplitude to 1, and fmax (Hz) to 700 Hz and keep the other parameters the same.

Step 4: In the Block Diagram window, right click on the screen. Under the Functions dialog box, click on Signal Processing and then select Embedded Signal Generation. Now click on Simulate Signal.

Step 5: Right click on Simulate Signal and click on Properties. Keep the frequency as 16 kHz, Amplitude as 1, Samples per second (Hz) as 32 kHz and No. of samples as 128.

Step 6: In the Block Diagram window, right click on the screen. Under the Functions dialog box, click on Programming and select Numeric. Now click on Add.

Step 7: Right click on Add, select Numeric Palette and click on Numeric Constant. Place it in the block diagram and change it from 0 to 1.

Step 8: In the Block Diagram window, right click on the screen. Under the Functions dialog box, click on Programming and select Numeric. Now click on Multiply.

Step 9: In the Block Diagram window, right click on the screen. Under the Functions dialog box, click on Signal Processing and select Frequency Domain. Now click on Spectral Measurements.

Step 10: Right click Spectral Measurements and click on Properties. After that, the Configure Spectral Measurements dialog box appears, then change Spectral Measurement to Magnitude (peak) and click on OK.

Step 11: In the Front Panel, add four Waveform Graphs and change the Labels to Input Signal, Input Signal—Frequency, Modulated Signal (AM) and Modulated Signal (AM)—Frequency as shown in Figure 3.20.

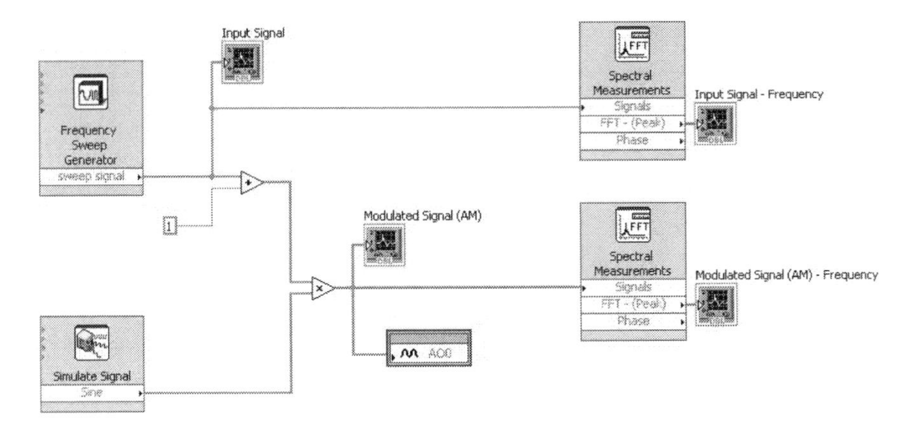

FIGURE 3.20 Block Diagram of AM modulator.

Step 12: Add an Analog Output.vi (AO0) and connect it to the output of the multiply block.

Step 13: Run the VI and notice the output in the front panel for different input signals as shown in Figure 3.21.

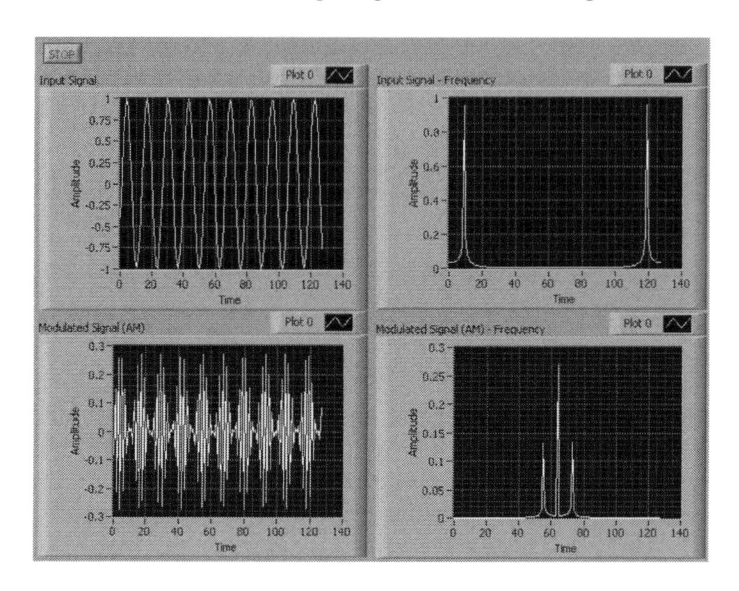

FIGURE 3.21 AM output and FFT output.

Step 14: Connect the demodulator of AM as shown in Figure 3.22.

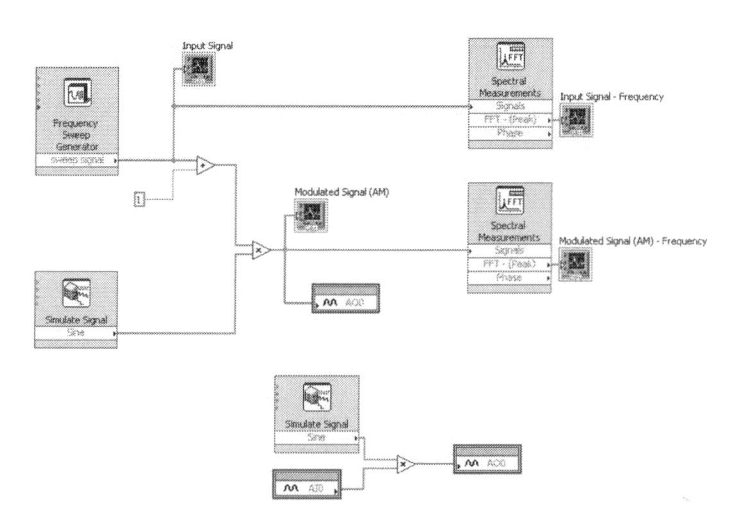

FIGURE 3.22 Final Block Diagram of AM demodulator.

3.6 Echo

An echo is a reflection of the sound waves from any surface back to the user. In this section, an echo is generated using the SPEEDY-33. Here an audio signal is fed to the board and is played back to the user after some delay which can range between milliseconds and several seconds. An echo with feedback sends the output of the circuit back to the input thus generating a continuous echo [3].

Implementation Steps

Step 1: Right click on the Block Diagram and then select Functions. Now select Programming, Elemental I/O and click on Analog Input.vi.

Step 2: Right click on AIO and select Properties. Keep the Sample Rate in Hz as 8000, change the Framesize to 1 and Gain as 1.

Step 3: Right click on the Block Diagram and then select Functions. Now select Programming, Elemental I/O and click on Analog Output.vi. Right click on AIO and select Properties. Keep the Sample Rate in Hz as 8000.

Step 4: Right click on the Block Diagram and then select Functions. Now select Programming, Numeric and click on Add.

Step 5: Right click on the Block Diagram and then select Functions. Now select Programming, Array and click on Index Array.

Step 6: Right click on the Block Diagram and then select Functions. Now select Programming, Array and click on Replace Array Subset.

Step 7: Right click on the Front Panel and then select Controls. Now select Modern, Numeric and click on Horizontal Fill Slide.

Step 8: Right click on Horizontal Fill Slide and select Properties. The Properties Configuration Box will appear on the screen. Change the label to Gain. Go to the Scale Tab and change Maximum to 1. Click on OK.

Step 9: In the Block Diagram window, right click on the screen. Under the Functions dialog box, click on Programming and select Numeric. Now click on Multiply.

Step 10: In the Block Diagram window, right click on the screen. Under the Functions dialog box, click on Programming and then select Numeric. Now click on Quotient & Remainder.

Step 11: In the Front Panel window, right click on the screen. Under the Controls dialog box, click on Modern and then select Numeric. Now click on Horizontal Pointer Slide. Rename the label as Delay in the block diagram. Right click on it and select Properties. Go to the Scale tab and change Maximum to 1000. Click on OK.

Step 12: In the Block Diagram window, right click on the screen. Under the Functions dialog box, click on Programming and then select Numeric. Now click on Increment in the Block Diagram.

Step 13: In the Block Diagram window, right click on the screen. Under the Functions dialog box, click on Programming and then select Structures. Now click on While Loop. Click and drag the While Loop from the top left corner to the bottom right corner to include all the blocks.

Step 14: Right click on the While Loop and click on Add Shift Register in the block diagram on either side of the While Loop. Add a 2nd Shift Register to the While Loop.

Step 15: In the Block Diagram window, right click on the screen. Under the Functions dialog box, click on Programming and then select Embedded Signal Generation. Now click on Simulate Signal.

Step 16: Right click on the Simulate Signal; a Configure Simulate Signal dialog box appears here. Change the Signal Type to DC and the Number of samples to 1000.

Step 17: Connect the Simulate Signal to the first Shift Register.

Step 18: Right click on the second Shift Register and select Create and then click on Constant.

Step 19: Connect all the components as shown in final Echo block diagram shown in Figure 3.23.

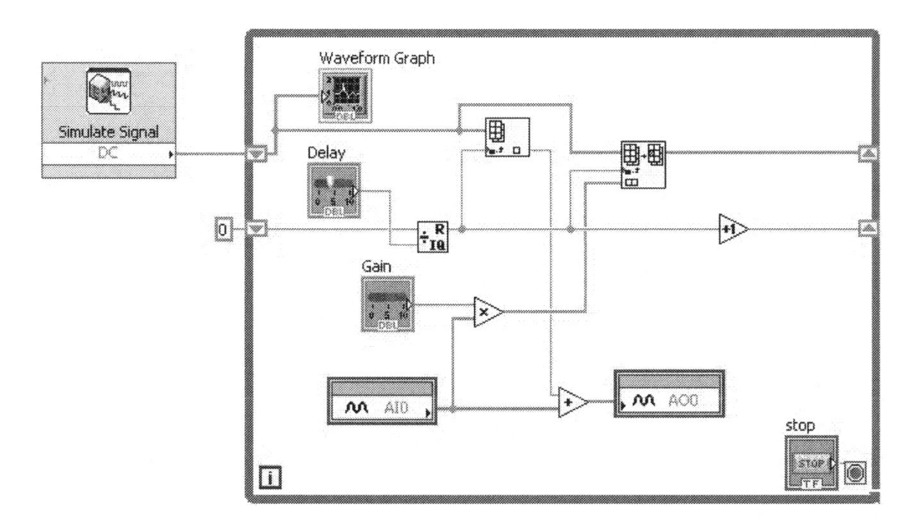

FIGURE 3.23 Final Block Diagram of Echo.

Step 20: To view results on a graph, right click on the Front Panel and then select Controls. Now select Modern, Graph and click on Waveform Graph as shown in Figure 3.24.

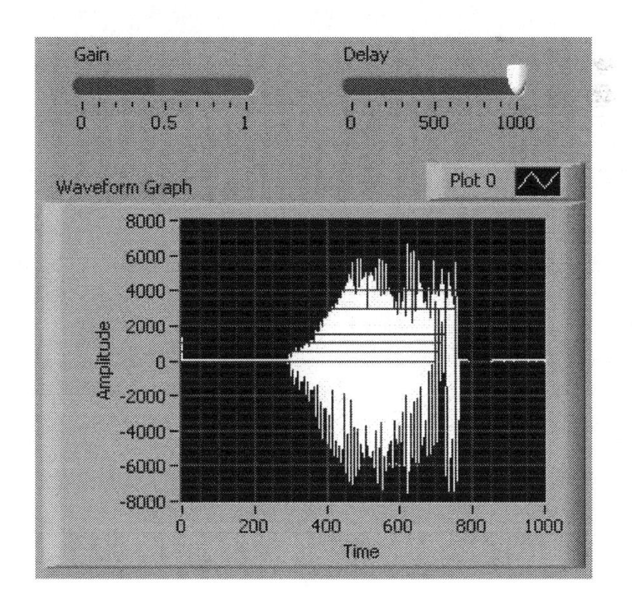

FIGURE 3.24 Echo output.

Step 21: In order to connect the Echo with Feedback, add another Add block and connect it as shown in Figure 3.25.

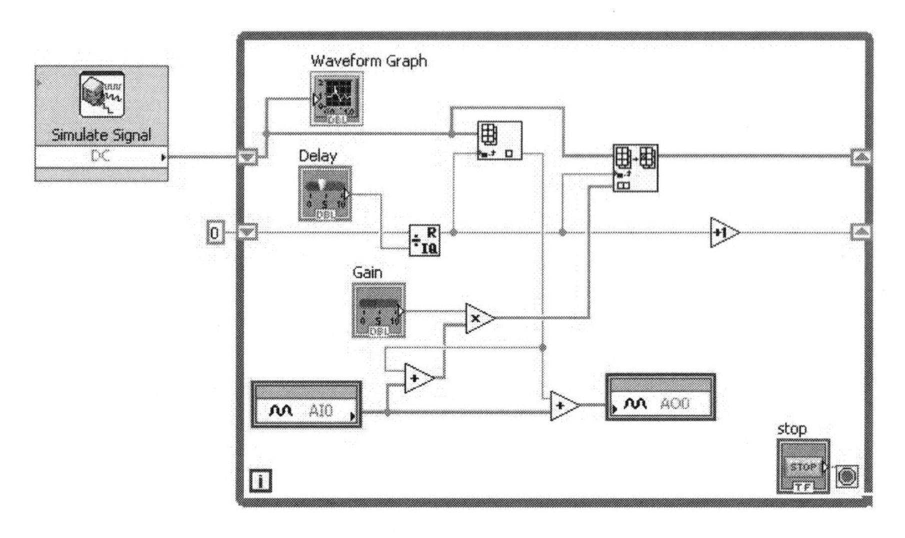

FIGURE 3.25 Echo with feedback.

3.7 Reverberation

Reverberation is the combination of the original signal with a delayed version of less than 20 ms. In this section, a reverberation is generated using the SPEEDY-33. Here an audio signal is fed to the board and is played back to the user after a very short delay of less than 20 ms. The final output sounds like a mixture of echos played closely together after the original signal is no longer heard.

Implementation Steps

Step 1: Follow the steps for creating a Reverb from Step 1 to Step 20 of the Echo in Section 3.6 and connect the blocks as shown in Figure 3.23. Change the Framesize of the AIO back to 128 and click on OK. Connect the input of the Multiply to the output of the Add as shown in Figure 3.26. Right click on the Numeric Constant and select Representation and change it from DBL (double precision) to I32 (long). Right click on the horizontal slide and select Representation and change it from DBL (double precision) to I32 (long) and change the label from Delay to Reverb.

Step 2: Now run the VI and observe the results.

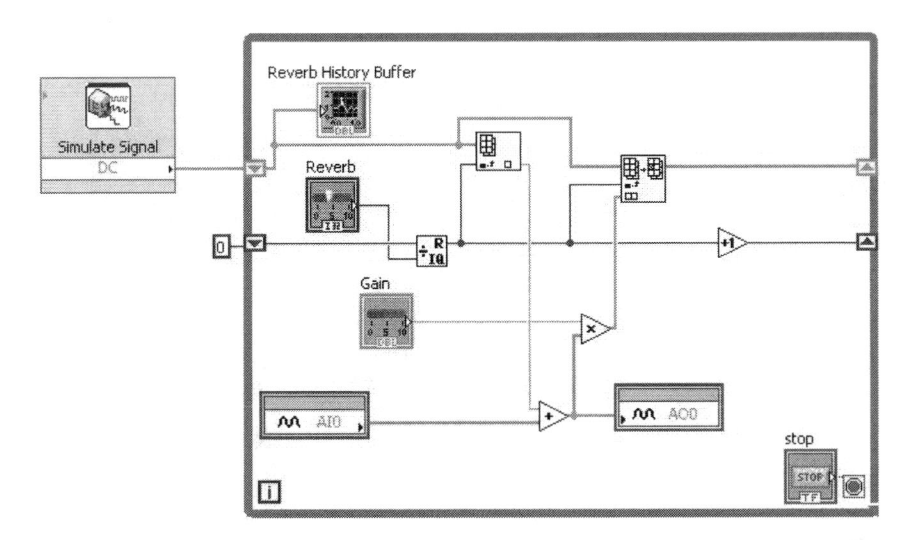

FIGURE 3.26 Block Diagram of Reverberation.

3.8 Creating Digital Music

In this section digital music is created using SPEEDY-33. The pitch is the frequency of the wave whereas the loudness is the amplitude of the wave. A number of frequencies are generated from A0 having a frequency of 27.5 Hz, to C8 having a frequency of 4186.01 Hz, which is the entire range of an acoustic piano. There are 12 notes in an octave and 88 piano keys on the keyboard. Every octave doubles the frequency of the previous octave. A chord is created when two or more notes are sounded harmonically together and are mixed arithmetically [3].

Implementation Steps

Step 1: In the Front Panel window, right click on the screen. Under the Controls dialog box, click on Modern and then select Boolean. Now click on the OK button.

Step 2: Right click on the OK button and click on Properties. Deselect Visible under Label, the Height to 180 to increase the length of the key, the width as 36 and the Off text as A for the first key then Click on OK.

Step 3: Add 6 more OK buttons and label each of the Off text from B to G as shown in Figure 3.27.

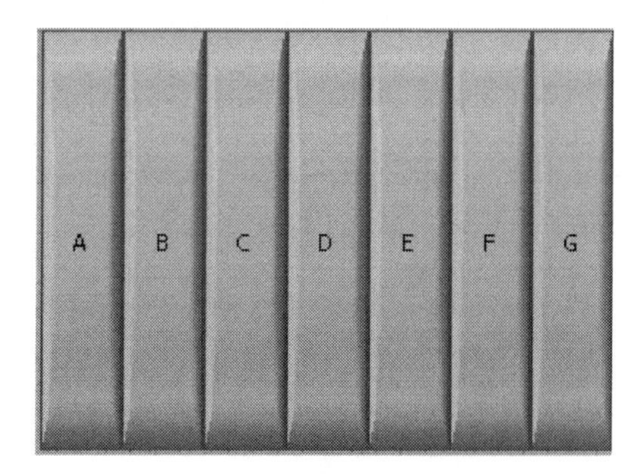

FIGURE 3.27 Keys "A" to "G."

Step 4: The 7 OK buttons appear in the Block Diagram.

Step 5: In the Block Diagram window, right click on the screen. Under the Functions dialog box, click on Programming and then select Embedded Signal Generation. Now click on EMB Sine Waveform.vi.

Step 6: One can also choose EMB Sine Waveform.vi by clicking on Programming under the Functions dialog box and then selecting Signal Processing, Embedded Signal Generation and then EMB Sine Waveform.vi.

Step 7: Add 7 EMB Sine Waveform.vi in the Block Diagram.

Step 8: Right click on the input of the EMB Sine Waveform.vi in the Block Diagram, select Create and click on Constant.

Step 9: Change the constant values to those shown in Figure 3.28.

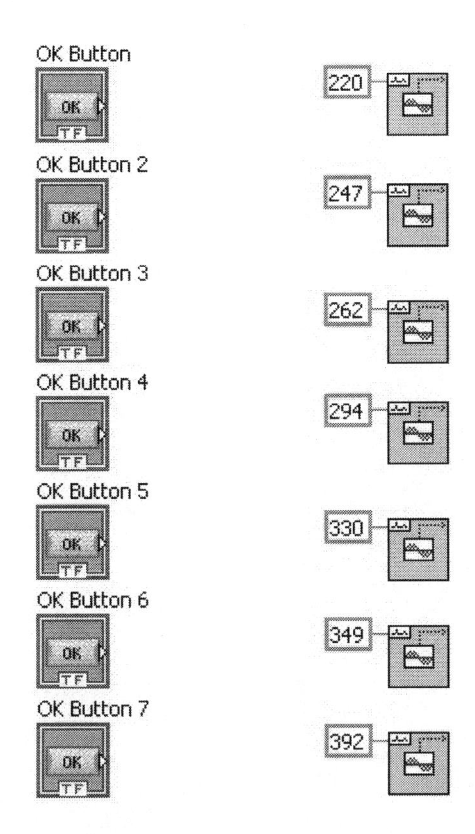

FIGURE 3.28 Embed sine wave constant values.

Step 10: In the Front Panel window, right click on the screen. Under the Controls dialog box, click on Modern and then select Numeric. Now click on Knob.

Step 11: Now right click on Knob, select Representation and click on DBL (Long) then place it in the diagram.

Step 12: Now right click on the Knob and select Properties. Go to the Tab "Scale" and change the Minimum to 0 and Maximum to 2000.

Step 13: In the Block Diagram window, right click on the screen. Under the Functions dialog box, click on Programming and then select Comparison.

Step 14: Place 7 comparison blocks in the Block Diagram.

Step 15: Connect the output of the Knob to all the inputs of the comparison select block. Now right click on the input of the comparison and click on Create and then Constant.

Make the value 0 (connect this constant to second input of all comparison blocks).

Step 16: In the Block Diagram window, right click on the screen. Under the Functions dialog box, click on Programming and select Numeric. Now click on Add.

Step 17: Place 6 Add Blocks and connect them as shown in Figure 3.29.

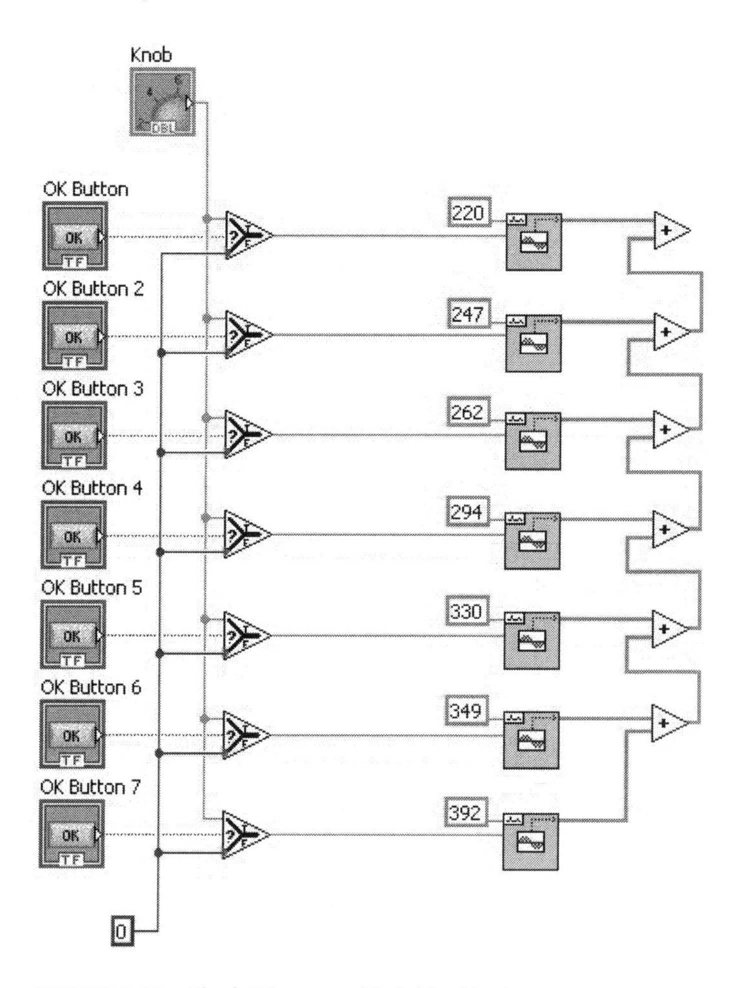

FIGURE 3.29 Block Diagram with Adder block.

Step 18: Right click on the Block Diagram and then select Functions. Now select Programming, Elemental I/O and click on Analog Output.vi.

Step 19: Place the Analog Output in the Block Diagram connecting the output of the topmost adder to both the analog outputs. Now drag the bottom edge of the Analog Output to display 2 Analog Outputs AO0 and AO1. Add 6 more OK buttons, EMB Sine Waveform.vi. and Add blocks and place them in the Front Panel as shown in Figure 3.30 and connect them in the Block Diagram as shown in Figure 3.31.

Step 20: The final Front Panel is shown in Figure 3.30.

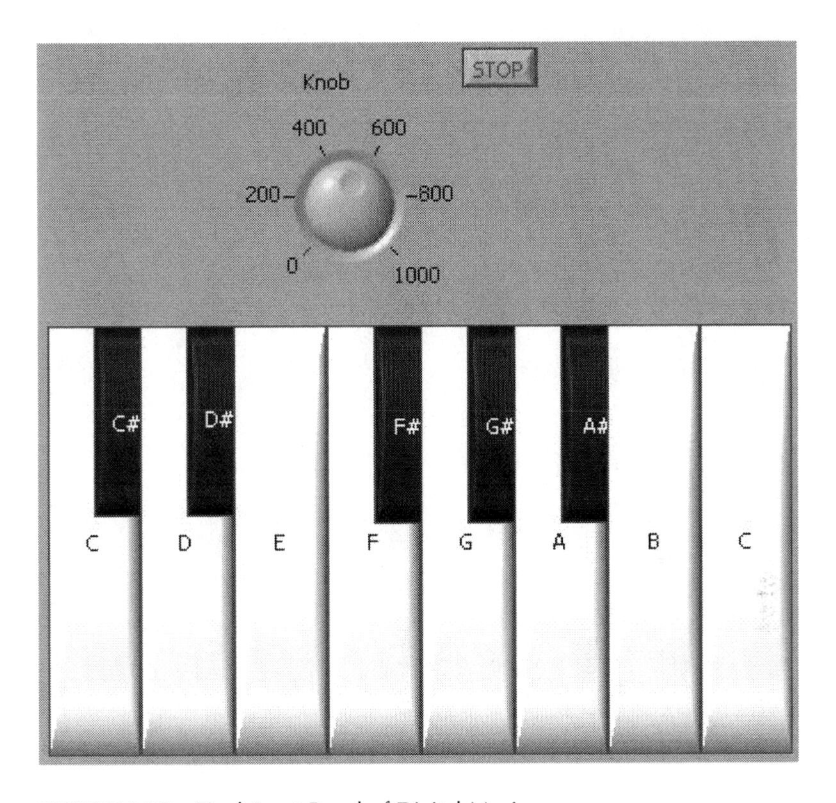

FIGURE 3.30 Final Front Panel of Digital Music.

Step 21: The final Block Diagram of digital music with while loop added is shown in Figure 3.31.

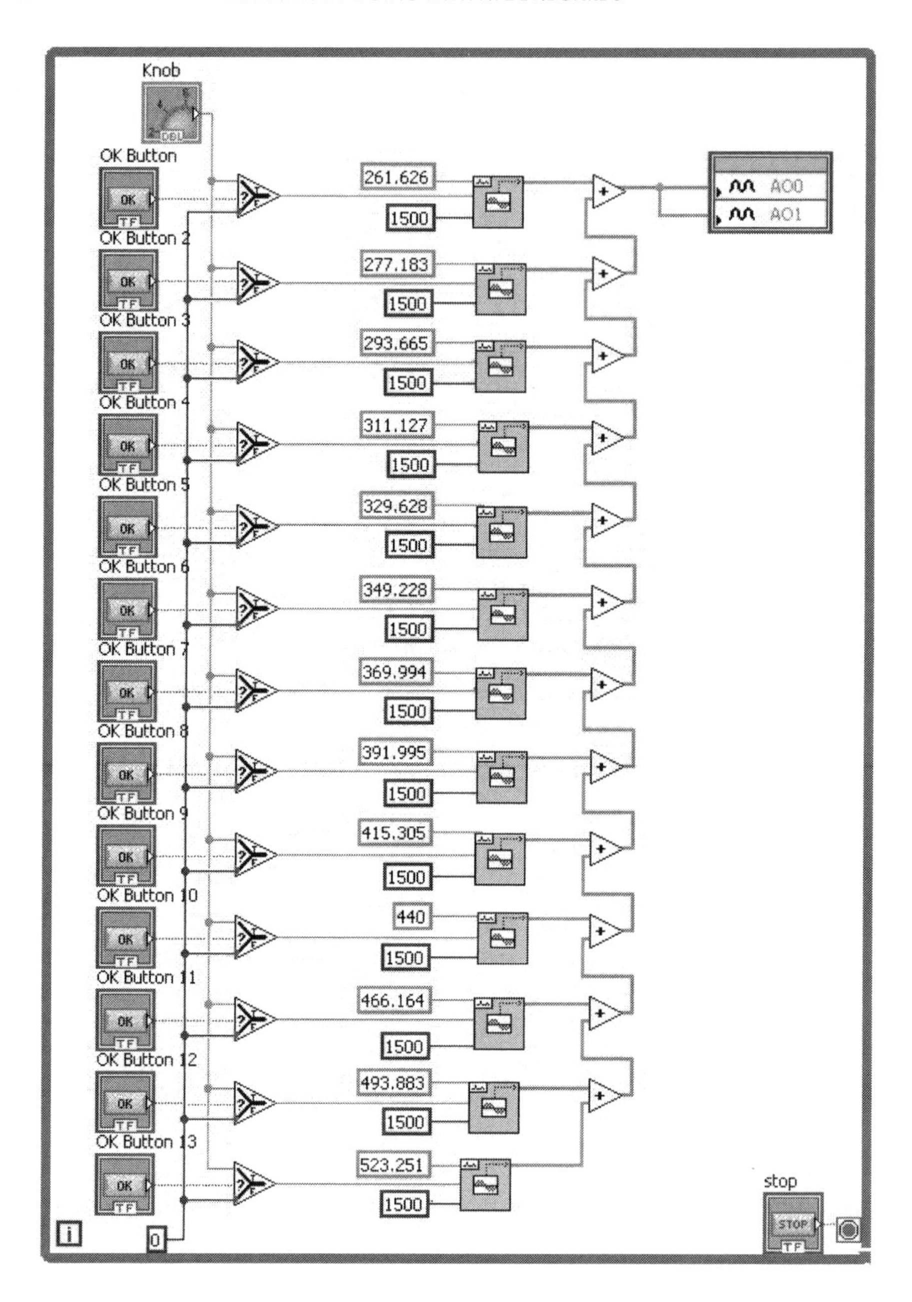

FIGURE 3.31 Final Block Diagram of Digital Music.

Step 22: The number of keys in Figure 3.31 can be extended to 88 keys by using the Note Names and Frequencies of Table 3.2.

Table 3.2 Note names and frequencies

	Note names											
Octave	C	C#/Db	D	D#/Eb	E	F	F#/Gb	G	G#/Ab	A	A#/Bb	B
0	–	–	–	–	–	–	–	–	–	27.5000	29.1353	29.1353
1	32.7032	34.6479	36.7081	38.8909	41.2035	43.6536	46.2493	48.9995	51.9130	55.0000	58.2705	61.7354
2	65.4064	69.2957	73.4162	77.7817	82.4069	87.3071	92.4986	97.9989	103.826	110.000	116.541	123.471
3	130.813	138.591	146.832	155.563	164.814	174.614	184.997	195.998	207.652	220.000	233.082	246.942
4	261.626	277.183	293.665	311.127	329.628	349.228	369.994	391.995	415.305	440.000	466.164	493.883
5	523.251	554.365	587.330	622.254	659.255	698.456	739.989	783.991	830.609	880.000	932.328	987.767
6	1046.50	1108.73	1174.66	1244.51	1318.51	1396.91	1479.98	1567.98	1661.22	1760.00	1864.66	1975.53
7	2093.00	2217.46	2349.32	2489.02	2637.02	2793.83	2959.96	3135.96	3322.44	3520.00	3729.31	3951.07
8	4186.01	–	–	–	–	–	–	–	–	–	–	–
	Frequency in Hz											

References

1. N. Kehtarnavaz and N. Kim. *Digital Signal Processing System-Level Design Using LabVIEW.* Elsevier/Newnes: Amsterdam/Boston, MA, 2005.
2. *Signal Processing Engineering Educational Device for Youth, NI SPEEDY-33 User Manual.* National Instruments: Austin, TX, 2008.
3. *User Guide and Specifications, NI myRIO-1900,* National Instruments: Austin, TX, May 2016.

CHAPTER FOUR

Exploring NI ELVIS

4.1 4-Bit Adder

A 4-Bit Adder is a circuit that adds two 4 binary numbers together to generate a 4-bit binary number (Out3 Out2 Outl Out0) plus a carry (Cout). It is also known as a Ripple Carry Adder since the carry bits from the output of the first adder ripple from one to the next and so on.

Implementation Steps

Step 1: The requirements for a 4-Bit Adder are 7408 IC, 7432 IC, 7486 IC, two DIP switches and five LEDs.

Step 2: Construct the circuit as shown in Figure 4.1.

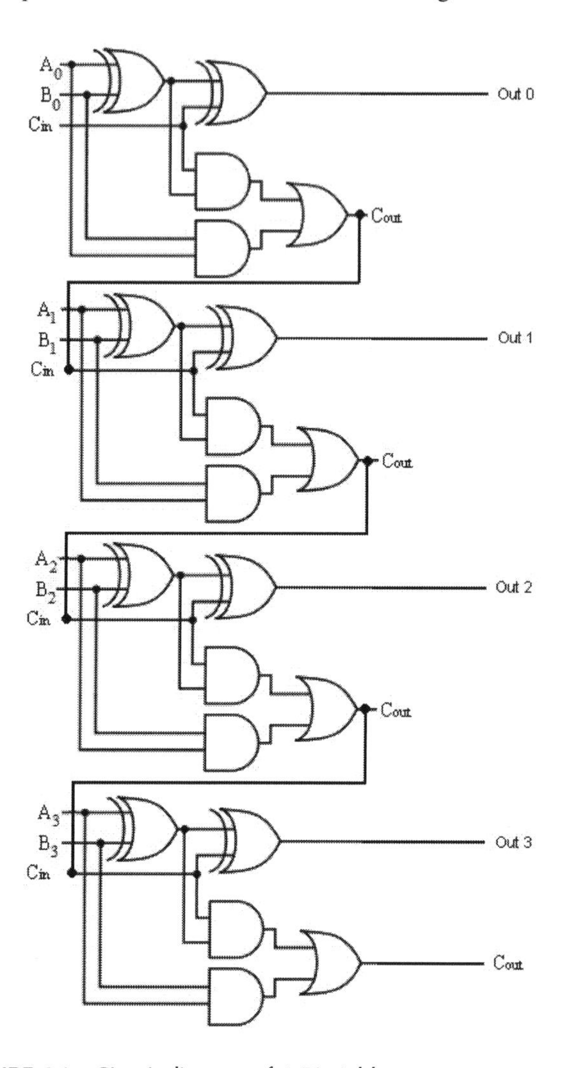

FIGURE 4.1 Circuit diagram of 4-Bit Adder.

Step 3: Next connect the Carry pin (Cin) to ground as shown in Figure 4.2.

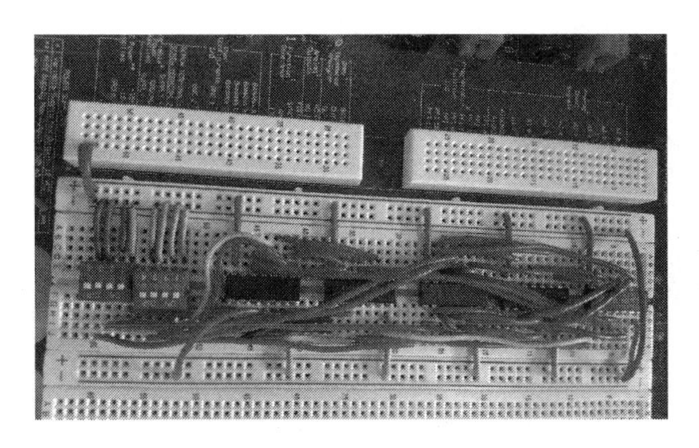

FIGURE 4.2 4-Bit Adder on Prototype Board.

Step 4: Finally, enter the inputs through the DIP switches and observe the output on the LEDs as shown in Figure 4.3. Here a bit "1" is entered through both DIP switches and "2," i.e. in binary "0010," is obtained at the output.

FIGURE 4.3 Output of Adder.

4.2 Traffic Light Control

The Traffic Light Control is used to provide coordination and control, such that traffic moves as safely and smoothly as possible. In this section, two 555 timers are used, which use a 10 V power supply [1]. The output of the left timer is through the red LED and the outputs of the right timer are through the yellow and green LEDs, wherein the output of the yellow LED is controlled by the discharge pin.

Implementation Steps

Step 1: The requirements for Traffic Light Control are two 555 timers, capacitors (2–16 V 100 μF), resistors (47 KΩ, 100 KΩ, 470 Ω and two 220 Ω) and LEDs (red, yellow and green).

Step 2: Construct the Traffic Light Control circuit as shown in Figure 4.4.

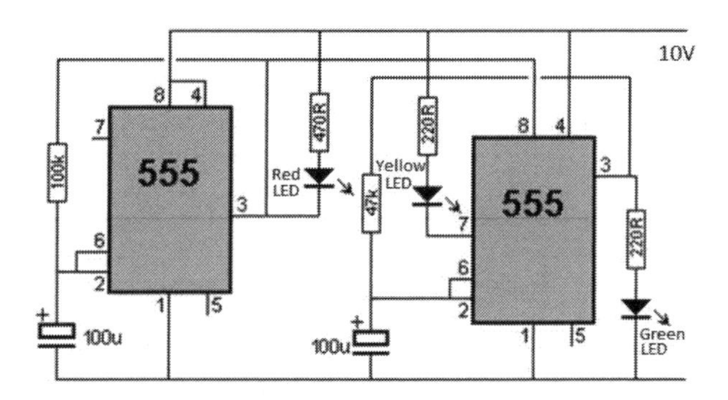

FIGURE 4.4 Traffic Light Control circuit diagram.

Step 3: Next, connect the power and ground wires to the variable power supply on the prototyping board.

Step 4: Now open the VI launcher and select the variable power supply option.

Step 5: Click on start button and set the voltage between 9 and 10 V. In order to vary using manual mode, click on manual button.

Step 6: Observe the outputs as shown in Figures 4.5–4.7. Green LED remains "ON" for 10 s, yellow LED remains "ON" for 4 s and red LED remains "ON" for 5 s; this happens continuously.

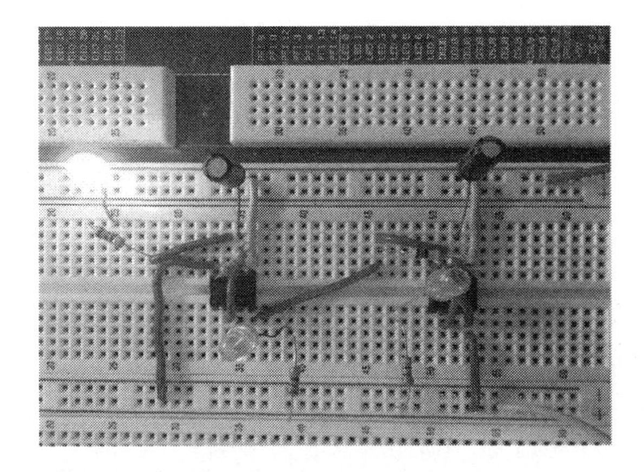

FIGURE 4.5 Green LED "ON" for 10 s.

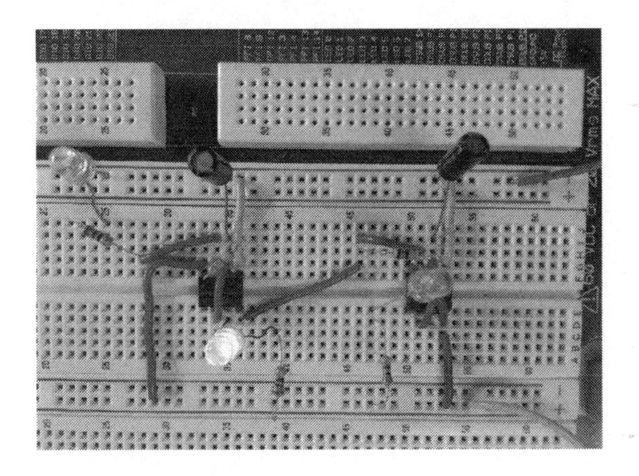

FIGURE 4.6 Yellow LED "ON" for 4 s.

FIGURE 4.7 Red LED "ON" for 5 s.

4.3 Digital Thermometer

Digital Thermometers are instruments that sense temperature, have permanent probes, are portable and have a digital display, therefore making them user friendly [2]. The working of a Digital Thermometer depends upon the sensor used, which could be a thermistor, a thermocouple or a resistance temperature detector. In this section, a LM35 sensor is used, which is an analog sensor and the electrical output is given in degrees centigrade. It displays the output on a DMM at 1 mV/C.

Implementation Steps

Step 1: The requirements for a Digital Thermometer are IC LM35, 10 KΩ and 2 KΩ pot.

Step 2: Construct the Digital Thermometer circuit as shown in Figure 4.8 on the prototyping board.

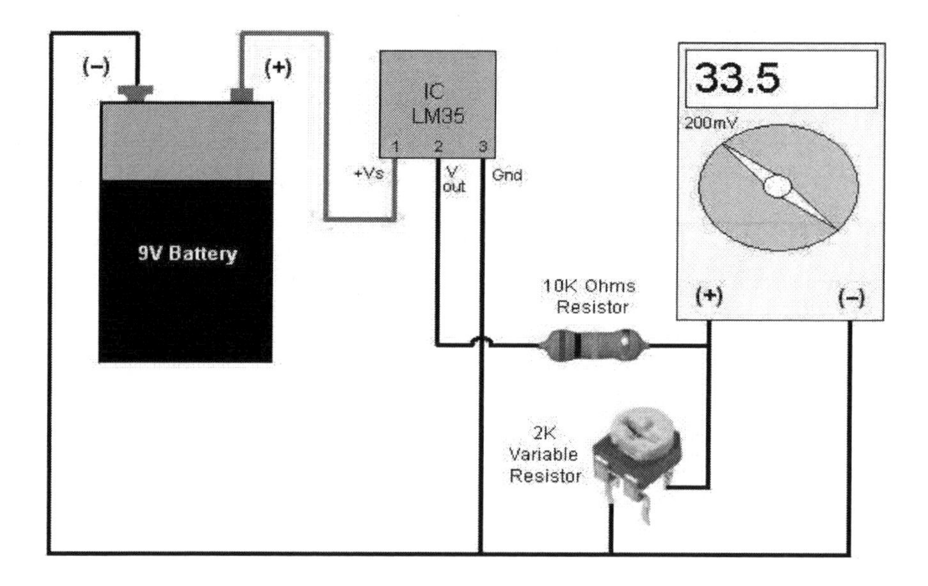

FIGURE 4.8 Circuit diagram of Digital Thermometer.

Step 3: Turn on the NI ELVIS power switch at the back corner.

Step 4: Connect the power and ground wires to the variable power supply on the prototyping board.

Step 5: Connect the USB to the PC.

Step 6: Open the VI launcher and select the variable power supply option.

Step 7: After building the circuit, set the voltage to 9 V in the variable power supply window and select 200 mV option on the DMM as shown in Figure 4.9.

FIGURE 4.9 Circuit diagram of Digital Thermometer on prototype board.

Step 8: Observe the output voltage on the DMM, which equals 1 mV/C.

4.4 Hearing Aid

A Hearing Aid is a small, portable electronic device that is to be worn behind or in the ear, in order to help the user hear well. It obtains sound through a mic that converts sound waves into electrical signals and finally amplifies them [3]. In this section, the audio is taken as the input and the amplified output is heard through the earphones.

Implementation Steps

Step 1: The requirements for a Hearing Aid are LM386, BC547, capacitors (100 μF, 4.7 μF, 470 μF, 22 μF, 47 μF, 0.1 μF (104)) and resistors (4.7 KΩ, 330 KΩ, 330 Ω, 150 Ω, 33 Ω, 10 Ω and 10 KΩ pot).

Step 2: Construct the circuit of a Hearing Aid as shown in Figure 4.10.

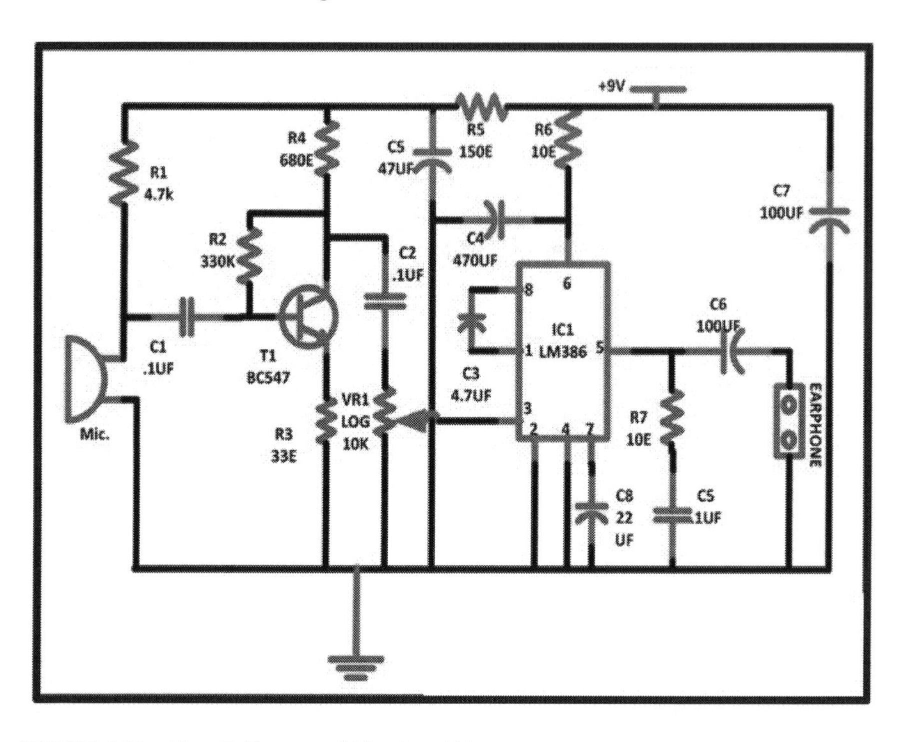

FIGURE 4.10 Circuit diagram of Hearing Aid.

Step 3: For power supply, connect the power and ground wires to the variable power supply on the prototyping board as shown in Figure 4.11.

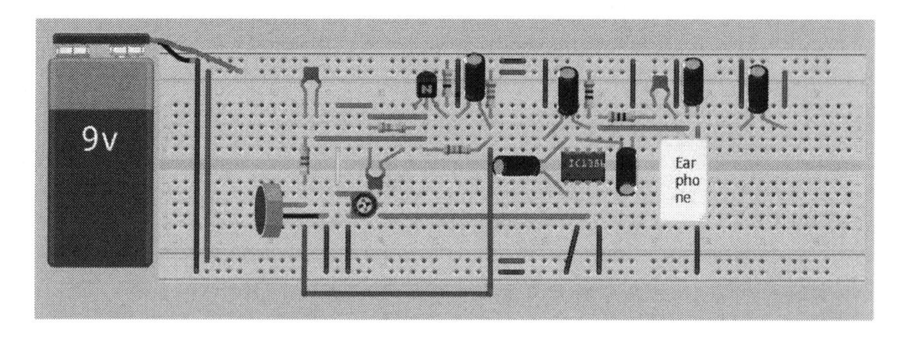

FIGURE 4.11 Connection of power supply to circuit.

Step 4: Open the VI launcher and select the variable power supply option.

Step 5: Now click on start button and set the voltage between 9 and 10 V.

Step 6: To vary using manual mode, click on manual button.

Step 7: To adjust the audio amplification, use the 10 KΩ pot as shown in Figure 4.12.

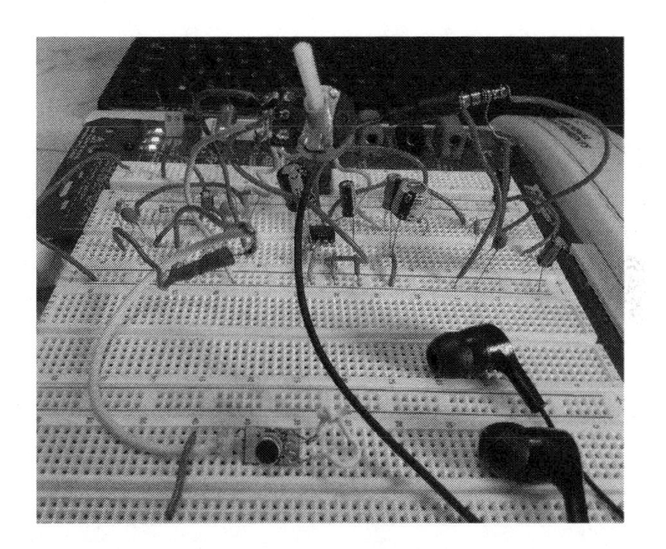

FIGURE 4.12 Implemented Hearing Aid circuit on prototype board.

Step 8: Hear the audio outputs using the earphones.

References

1. J. Butime, R. Besiga, A. Bwonyo, and A. Katumba. Design of online Digital Electronics laboratories based on the NI ELVIS II platform. In *9th International Conference on Remote Engineering and Virtual Instrumentation (REV)*, IEEE, 2012.
2. K.P.S. Rana, V. Kumar, and J. Mendiratta. An educational laboratory virtual instrumentation suite assisted experiment for studying fundamentals of series resistance–inductance–capacitance circuit. *European Journal of Engineering Education*, 42(3):1–20, 2017.
3. E. Doering. *Engineering Signals & Systems: Hands-On Labs with NI ELVIS*. National Technology & Science Press, 2019, ISBN-13: 978-1-934891-24-7.

CHAPTER FIVE

Exploring myRIO

5.1 4 × 4 Keypad Interfacing

A keypad interfaced to any embedded platform makes it easy for user to interact. Figure 5.1 shows the NI myRIO Systems interfaced to a 4 × 4 keypad. These are 4 × 4 push to on switches arranged in rows and columns. These keys are then scanned to identify which key is pressed [1].

In order to interface the keypad to myRIO, 8 Jumper Wires (M-F) are connected in the following way:

- P1: Row1 – pin 19 (B/DIO4)
- P2: Row2 – pin 21 (B/DIO5)
- P3: Row3 – pin 23 (B/DIO6)
- P4: Row4 – pin 25 (B/DIO7)
- P5: Column1 – pin 11 (B/DIO0)
- P6: Column2 – pin 13 (B/DIO1)
- P7: Column3 – pin 15 (B/DIO2)
- P8: Column4 – pin 17 (B/DIO3)

wherein P1 to P8 are the pins from left to right of the Keypad.

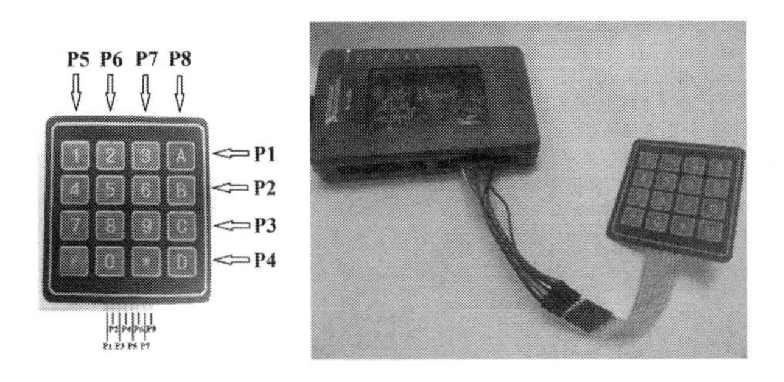

FIGURE 5.1 4 × 4 Keypad Interfacing.

Implementation Steps

Step 1: In the Block Diagram window, right click on the screen. Under the Functions dialog box, click on Programming and then select Structures followed by the For Loop as shown in Figure 5.2. Now click and drag from the left top corner to the bottom right corner to form a rectangle.

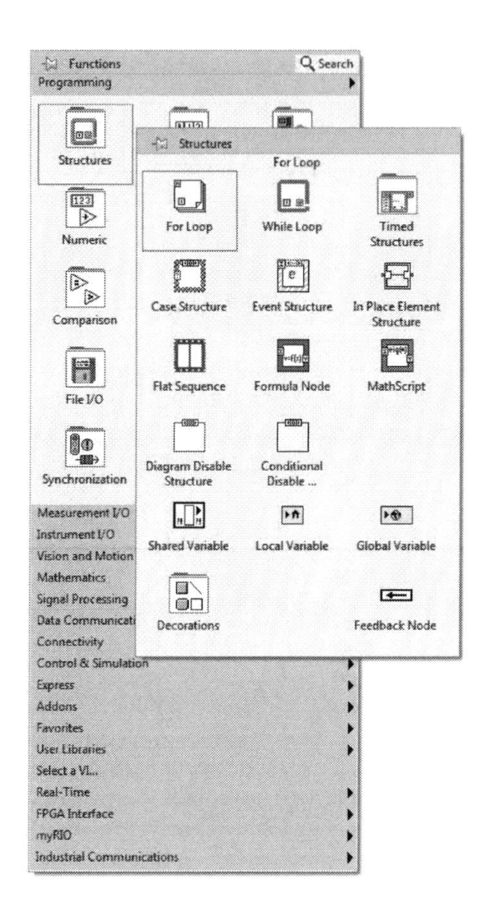

FIGURE 5.2 For Loop Selection.

Step 2: In the Block Diagram window, right click on the screen. Under the Functions dialog box, click on Programming and then select Structures followed by the While Loop as shown in Figure 5.3. Now click and drag from the left top corner to the bottom right corner to form a rectangle. See that the While Loop is the outermost loop which includes the For Loop.

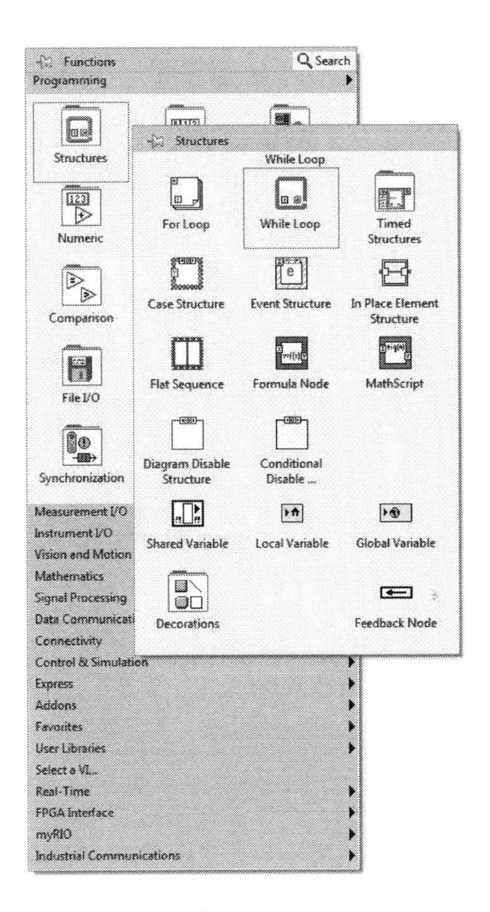

FIGURE 5.3 Selecting While Loop.

Step 3: In the Block Diagram window, right click on the screen. Under the Functions dialog box, click on Programming and then select Structures followed by the Case Structure as shown in Figure 5.4. Now click and drag from the left top corner to the bottom right corner to form a rectangle. See that the Case Structure is the innermost loop.

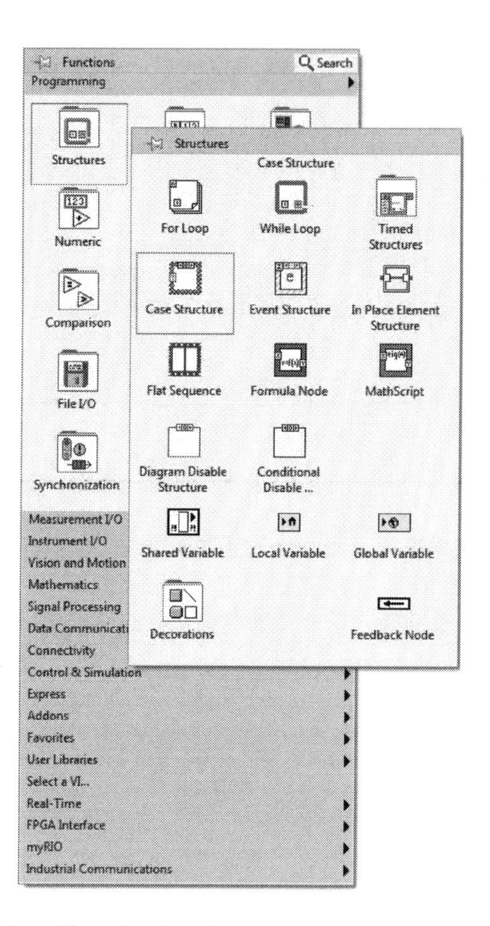

FIGURE 5.4 Choosing Case Structure.

Step 4: In the Block Diagram window, right click on the screen. Under the Functions dialog box, click on Programming and then select Array followed by the Build Array as shown in Figure 5.5. It will look like when it is placed in the block diagram. Change it to by right clicking on it and select Add Input four times. Add another Build Array and change it to .

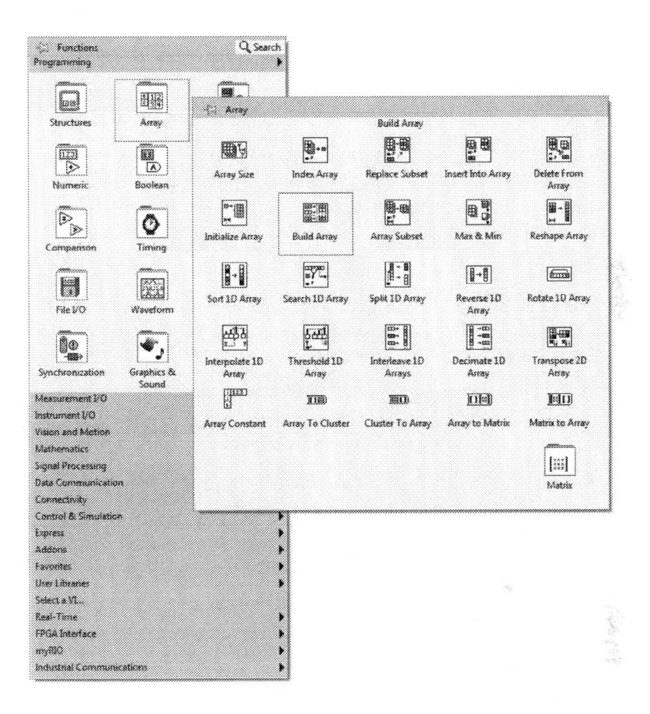

FIGURE 5.5 Build Array.

Step 5: In the Block Diagram window, right click on the screen. Under the Functions dialog box, click on Programming and then select Array followed by the Transpose 2D Array as shown in Figure 5.6.

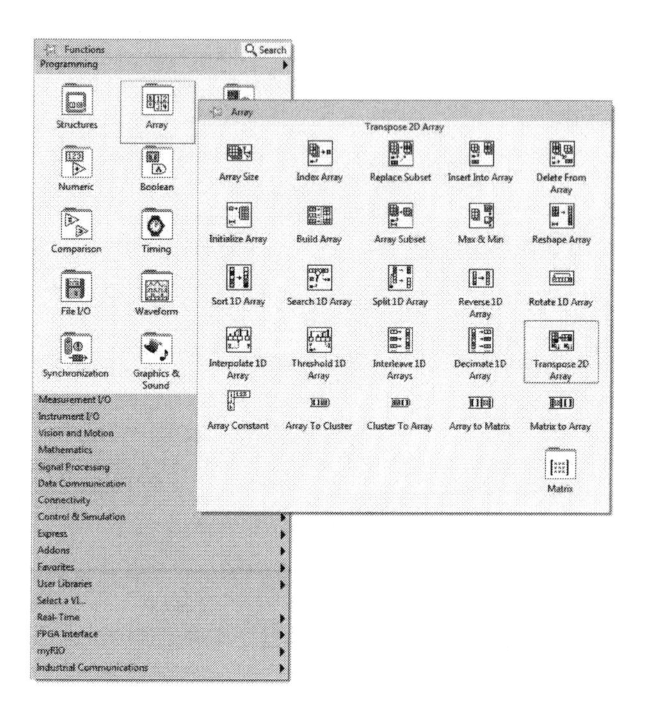

FIGURE 5.6 Transpose 2D Array.

Step 6: In the Block Diagram window, right click on the screen. Under the Functions dialog box, click on Programming and then select Numeric and then Numeric Constant as shown in Figure 5.7. Add two numeric constants. Change one of the values to 10 and the other to 4.

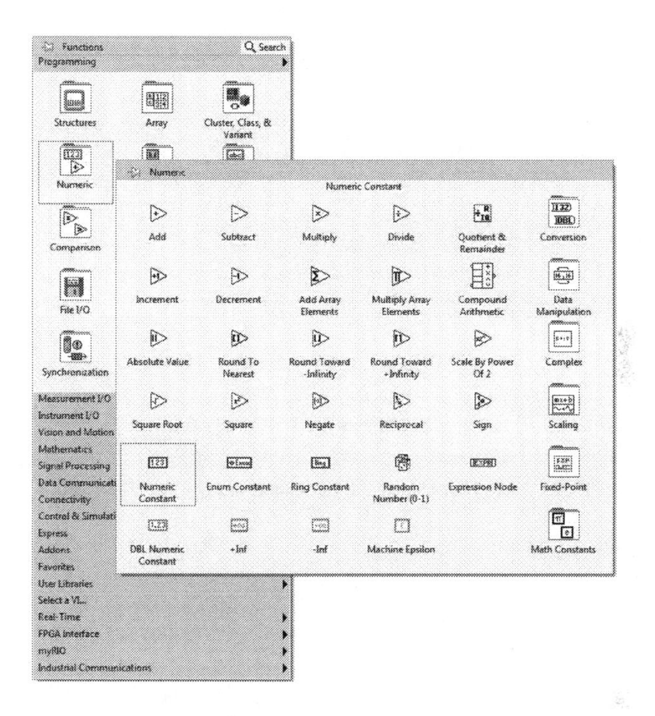

FIGURE 5.7 Numeric Constant.

Step 7: In the Block Diagram window, right click on the screen. Under the Functions dialog box, click on Programming and then select Boolean. Now click on Or as shown in Figure 5.8.

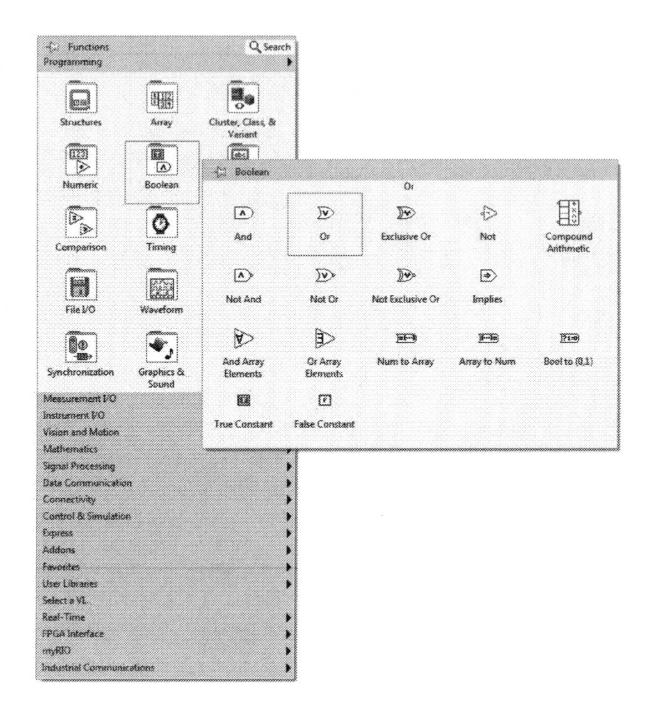

FIGURE 5.8 Selecting Or Gate.

Step 8: In the Block Diagram window, right click on the screen. Under the Functions dialog box, click on Programming and then select Boolean. Now click on Not as shown in Figure 5.9.

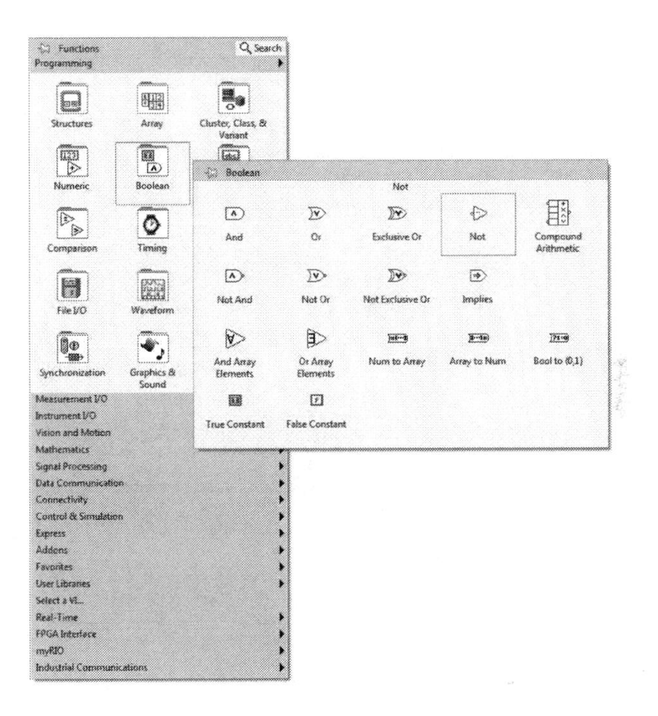

FIGURE 5.9 Selecting Not Gate

Step 9: In the Block Diagram window, right click on the screen. Under the Functions dialog box, click on Programming and then select Boolean followed by the False Constant as shown in Figure 5.10.

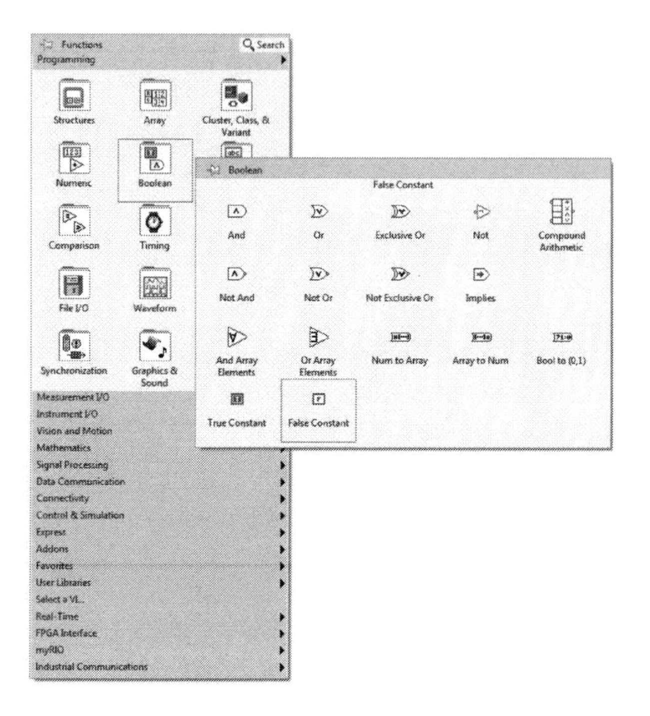

FIGURE 5.10 False Constant.

Step 10: In the Block Diagram window, right click on the screen. Under the Functions dialog box, click on Programming and then select Timing. Now click on Wait (ms) as shown in Figure 5.11.

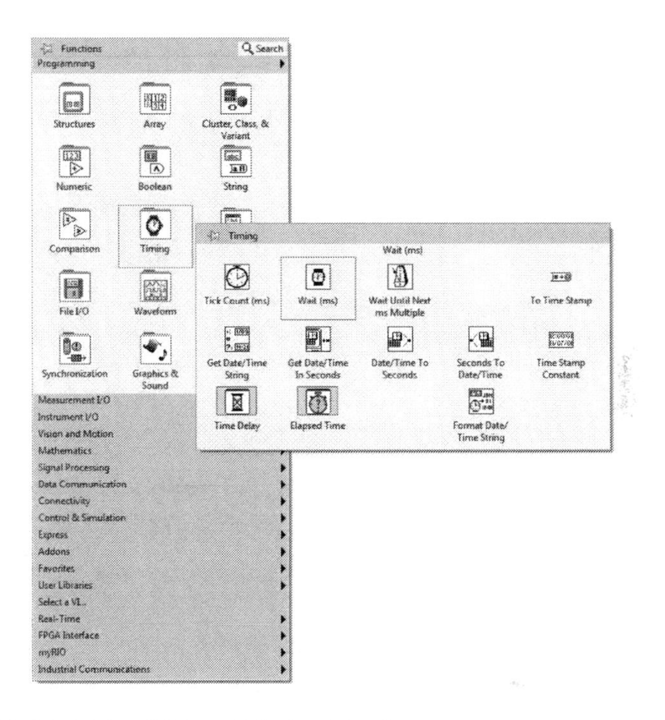

FIGURE 5.11 Wait (ms) Timing Selection.

Step 11: In the Block Diagram window, right click on the screen. Under the Functions dialog box, click on Programming and then select Dialog & User Interface. Now click on Simple Error Handler.vi as shown in Figure 5.12.

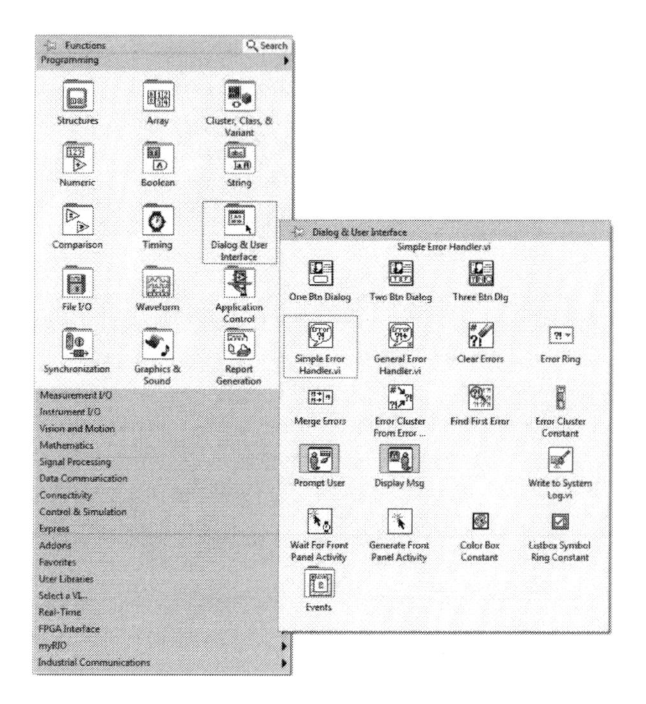

FIGURE 5.12 Simple Error Handler.vi.

Step 12: In the Block Diagram window, right click on the screen. Under the Functions dialog box, click on Programming and then select Array followed by the Merge Errors as shown in Figure 5.13. It will look like when it is placed in the block diagram.

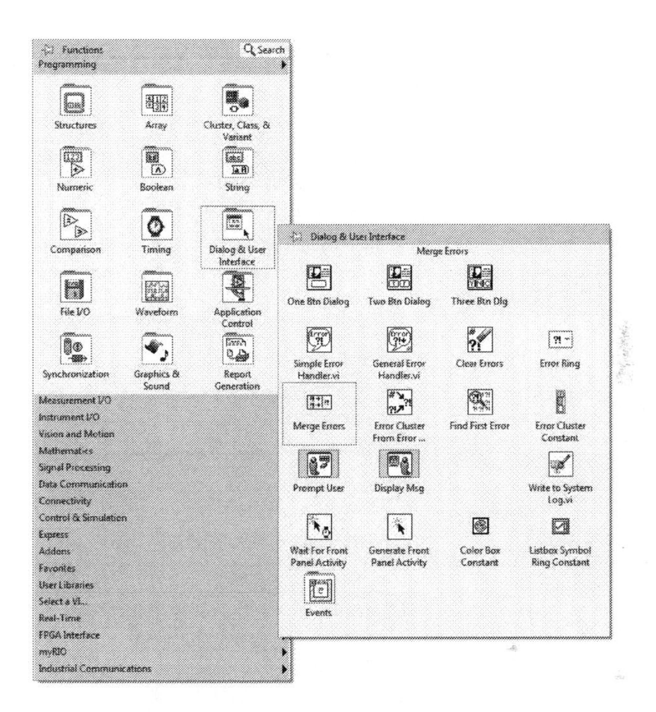

FIGURE 5.13 Merge Errors.

Step 13: In the Block Diagram window, right click on the screen. Under the Functions dialog box, click on myRIO and then select Digital In as shown in Figure 5.14.

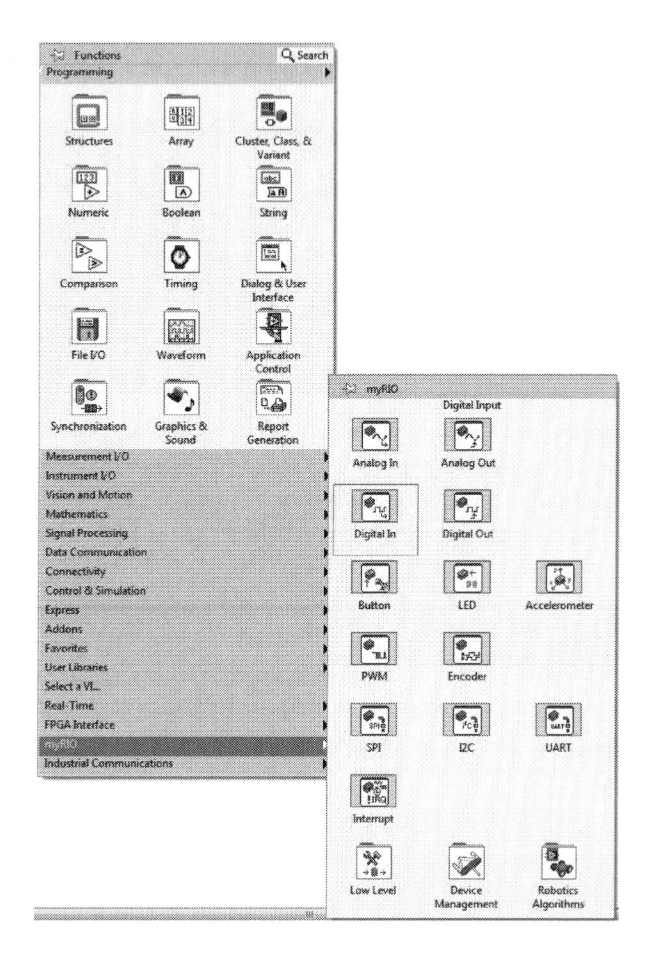

FIGURE 5.14 Digital In.

Step 14: A Configure Digital Input (Default Personality) dialog box appears. Change the channel to B/DIO4 (Pin 19). Click on the + button to display B/DIO5 (Pin 21). Click on the + twice more to display B/DIO6 (Pin 23) and B/DIO7 (Pin 25). It will look like Figure 5.15.

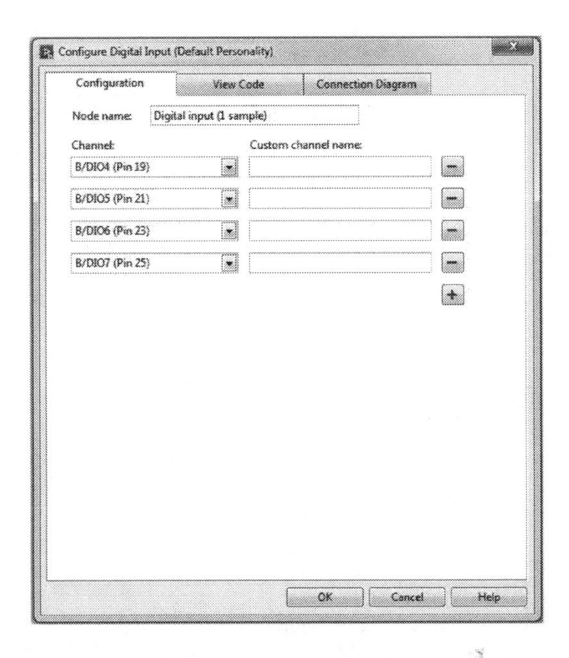

FIGURE 5.15 Configure Digital Input (Default Personality) Dialog Box.

Step 15: In the Block Diagram window, right click on the screen. Under the Functions dialog box, click on myRIO and then select Low Level and Digital I/O 1. Now click on Open as shown in Figure 5.16. After placing it in the Block Diagram, right click on one of the input (channel names) and select Create Constant. It will look like . Change the channel name by double clicking after I/O and typing in B/DIO0. Right click on the channel name and select Size To Text. Now right click on the dimension on the left side of the channel name and select Properties. Unselect the Show index display. Click on OK. It will look like . Similarly change the other inputs to B/DIO1, B/DIO2 and B/DIO3.

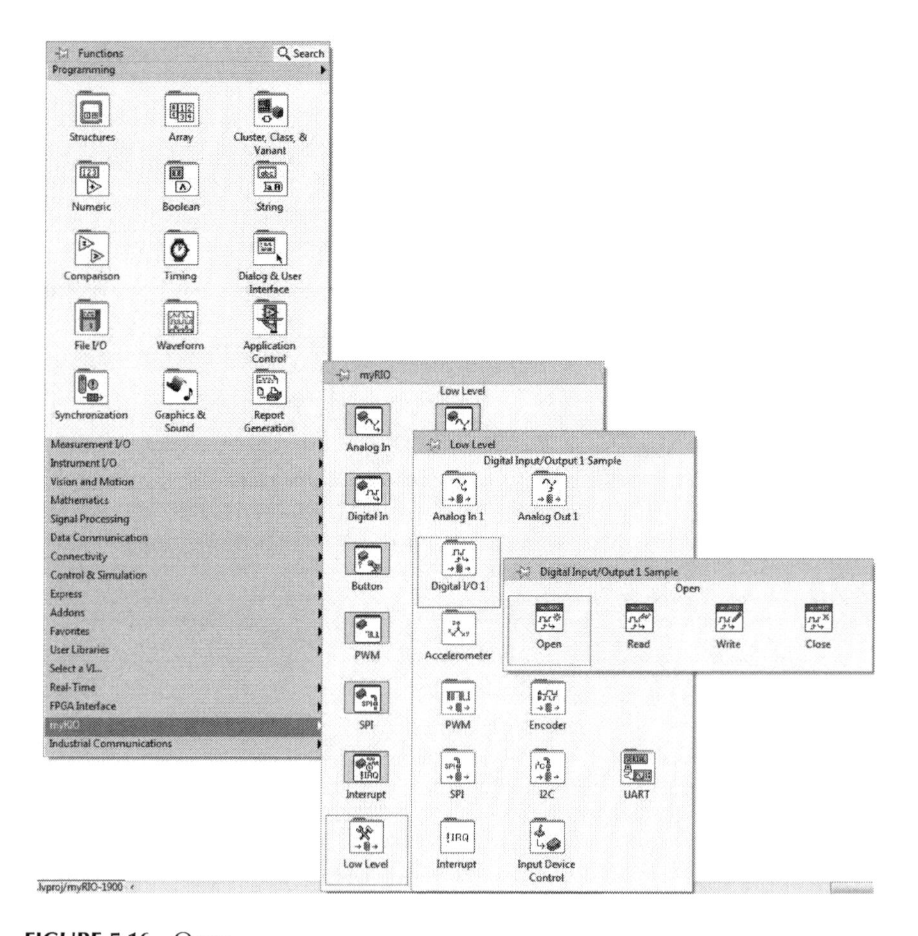

FIGURE 5.16 Open.

Step 16: In the Block Diagram window, right click on the screen. Under the Functions dialog box, click on myRIO and then select Low Level and Digital I/O 1. Now click on Read as shown in Figure 5.17.

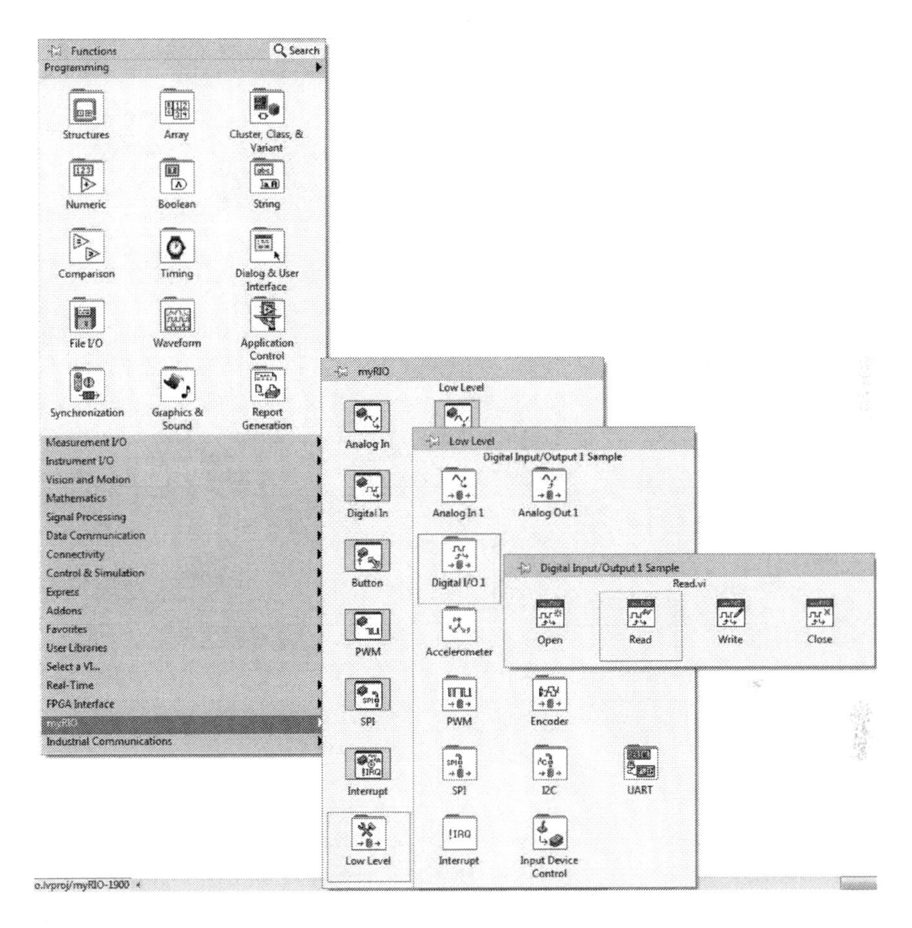

FIGURE 5.17 Read.

Step 17: In the Block Diagram window, right click on the screen. Under the Functions dialog box, click on myRIO and then select Low Level and Digital I/O 1. Now click on Write as shown in Figure. 5.18.

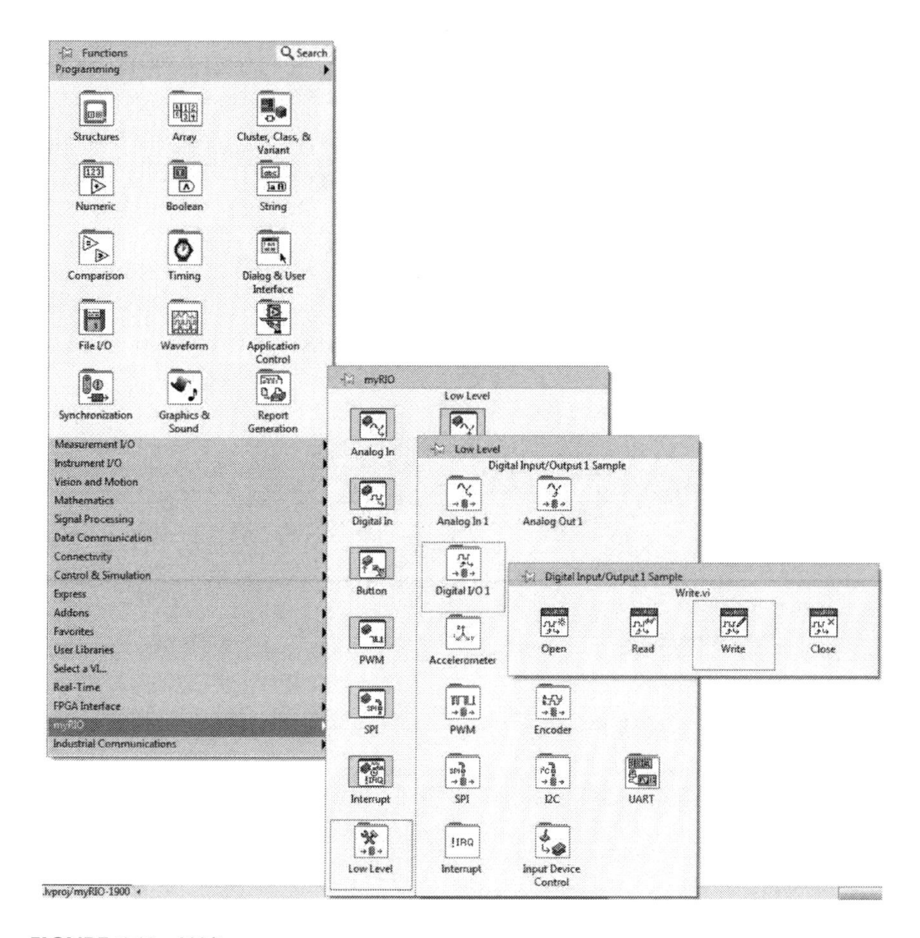

FIGURE 5.18 Write.

Step 18: In the Block Diagram window, right click on the screen. Under the Functions dialog box, click on myRIO and then Device Management. Now select Reset as shown in Figure 5.19.

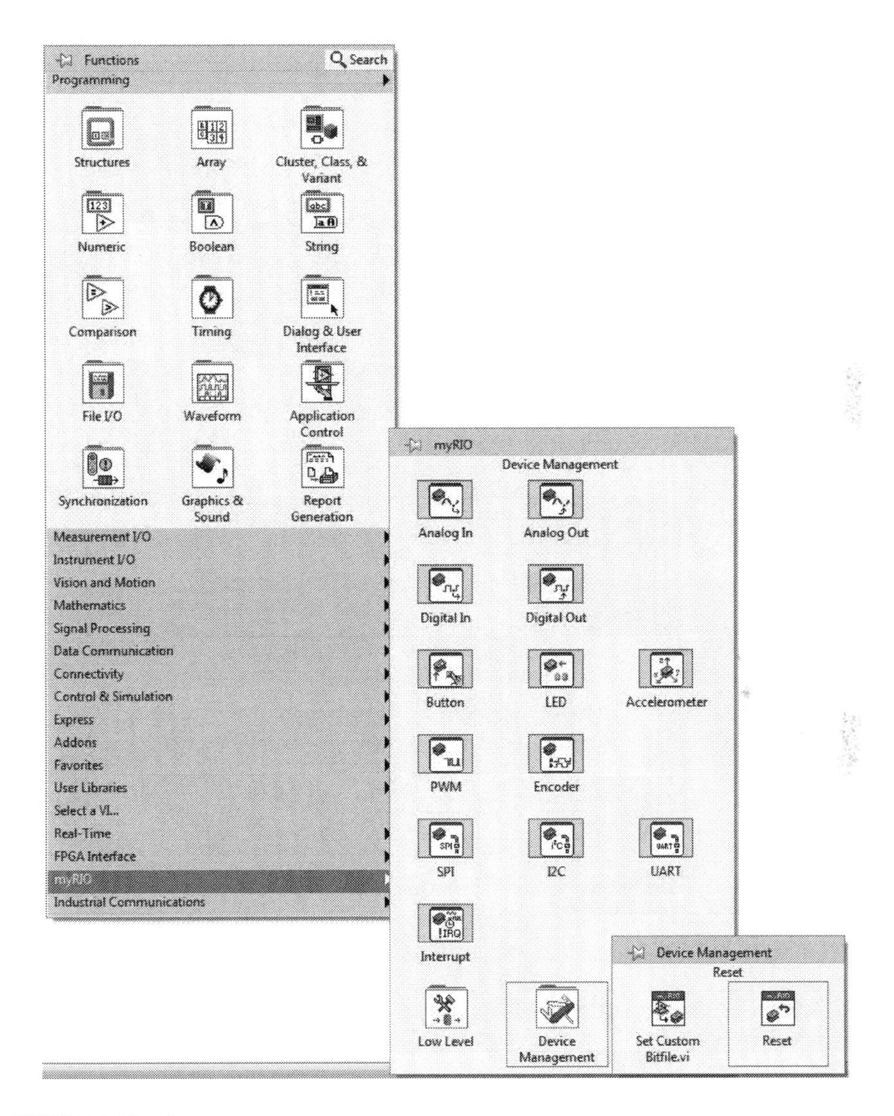

FIGURE 5.19 Reset.

Step 19: In the Front Panel window, right click on the screen. Under the Controls dialog box, click on Silver and then Boolean. Now select LED (Silver) as shown in Figure 5.20. Now select another LED (Silver) and place it in the front panel. Change one of the dimensions of the LED (Silver) by right clicking on it and selecting properties and then change the height and width to 10 and 10, respectively.

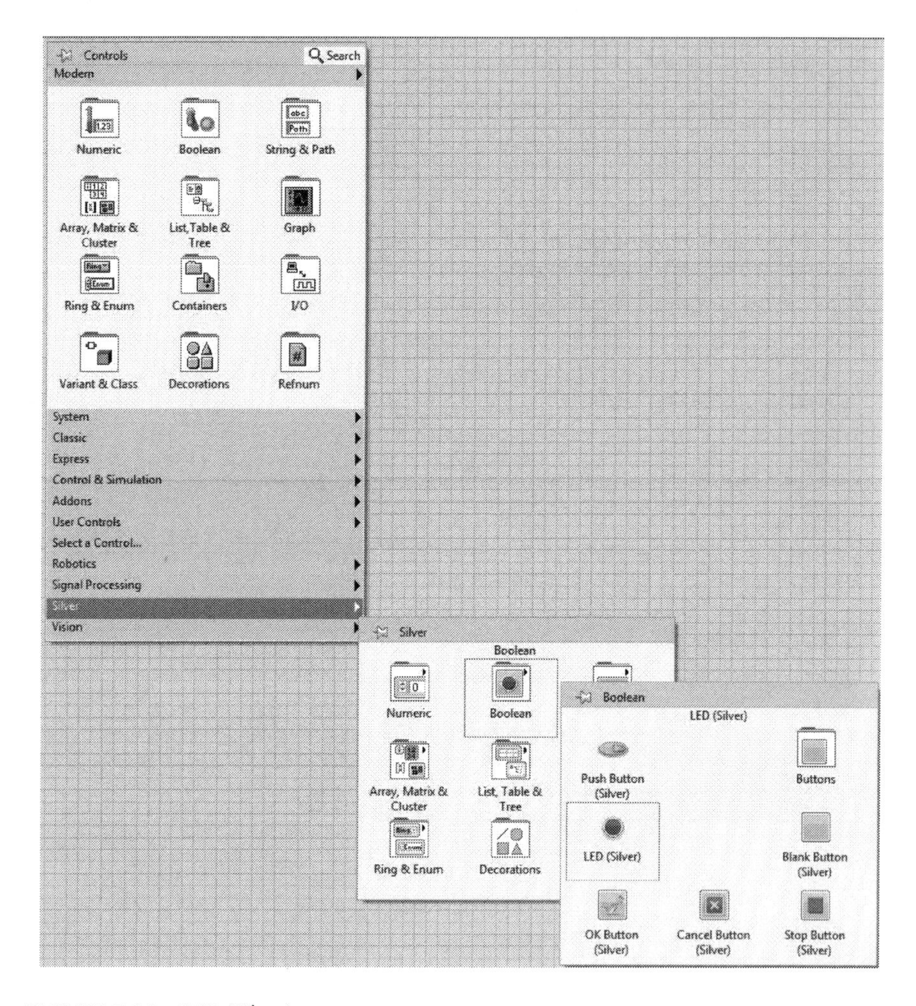

FIGURE 5.20 LED (Silver).

Step 20: In the Front Panel window, right click on the screen. Under the Controls dialog box, click on Silver and then Boolean. Now select Stop Button (Silver) as shown in Figure 5.21. Rename the label [ESC].

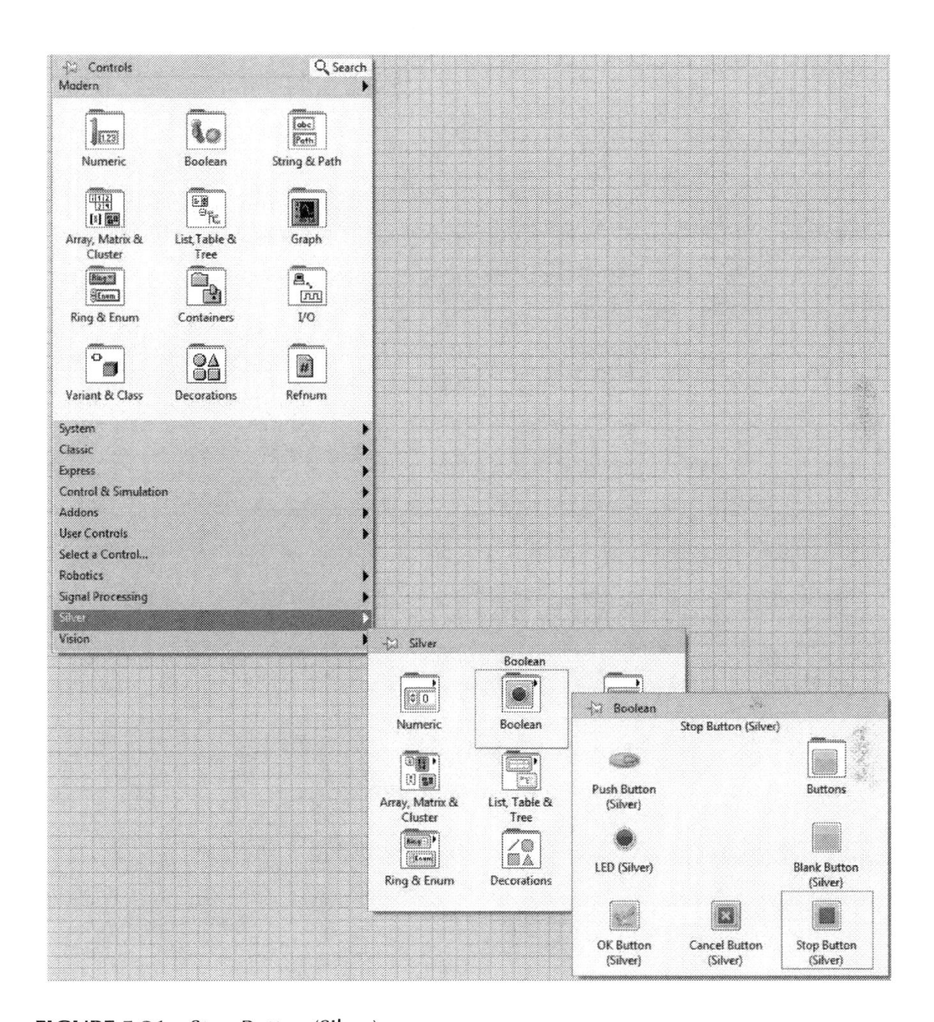

FIGURE 5.21 Stop Button (Silver).

Step 21: In the Front Panel window, right click on the screen. Under the Controls dialog box, click on Silver and then Array, Matrix & Cluster. Now select Array (Silver) as shown in Figure 5.22 and rename it Keypad Buttons. Select another Array (Silver) and place it in the front panel and rename it Scan.

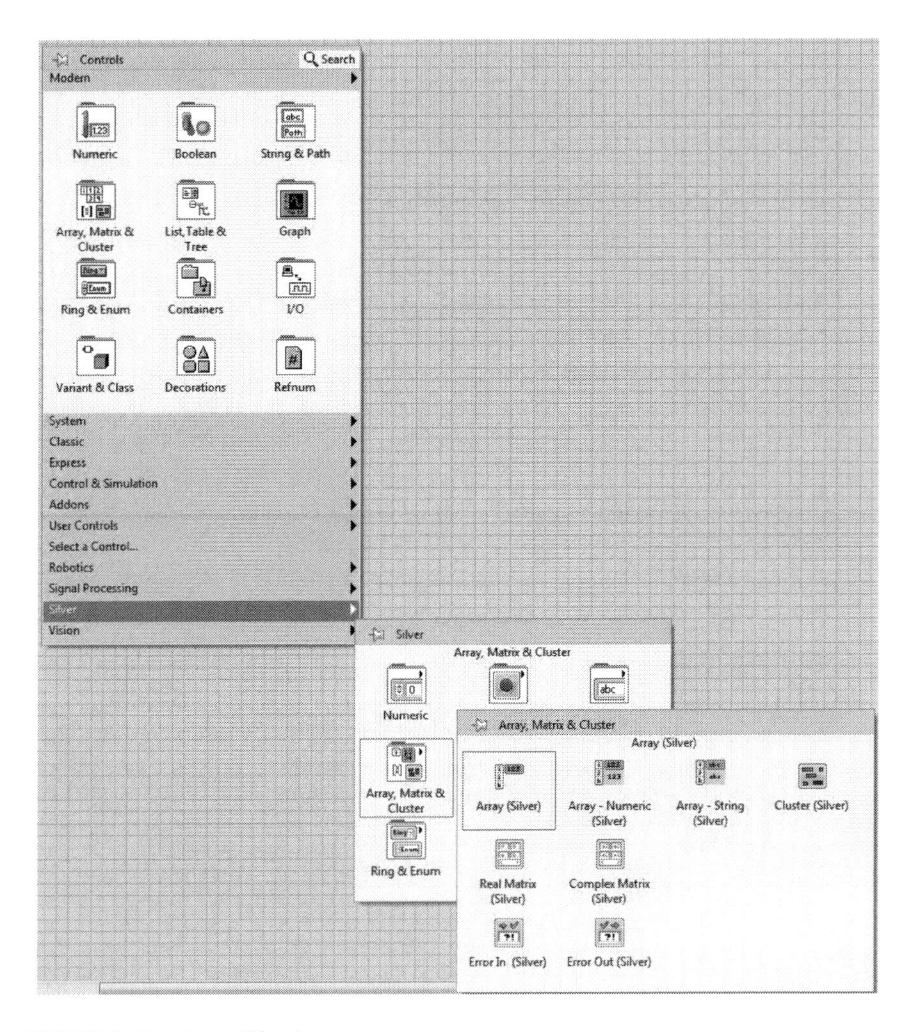

FIGURE 5.22 Array (Silver).

Step 22: Now drag one of the LED (Silver) of size 19 × 19 into the Array Keypad buttons. Right click on the array and select Add Dimension. Extend the array by dragging the right edge such that four LEDs are visible and then extend it further by dragging the bottom edge such that a matrix of 4 × 4 of 16 LEDs is visible. Now right click on the array and select Properties. Unselect the Show Index display. Click on OK. It will look like .

Step 23: Now drag one of the LED (Silver) of size 10 × 10 into the Array Keypad buttons. Extend the array by dragging the right edge in the right direction such that 16 LEDs are visible. Now right click on the array and select Properties. Unselect the Show index display. Click on OK. It will look like .

Step 24: Connect all the blocks as shown in the final view of the block diagram window in Figure 5.23. Right click on the For Loop at the connection point of the Open and Read, select Replace with Shift Register. Right click on the connection between the Not gate and the Scan array, select Tunnel Mode and then Concatenating.

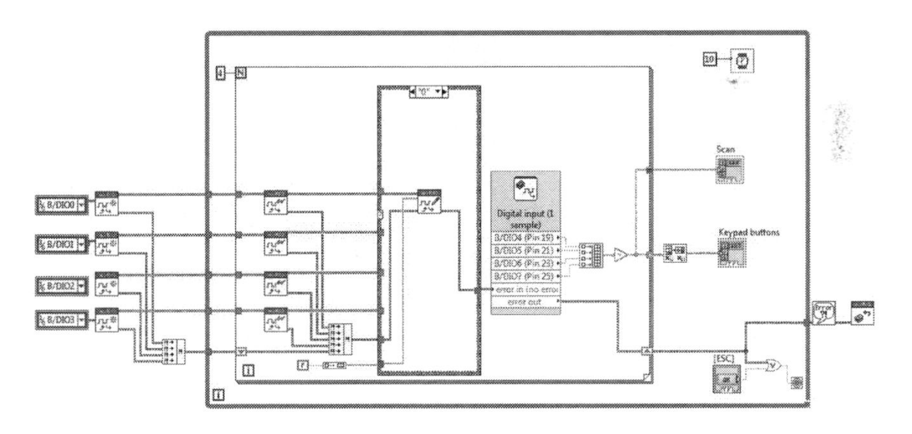

FIGURE 5.23 Final view of Keypad Block Diagram.

Step 25: The final view of the Front Panel window is shown in Figure 5.24.

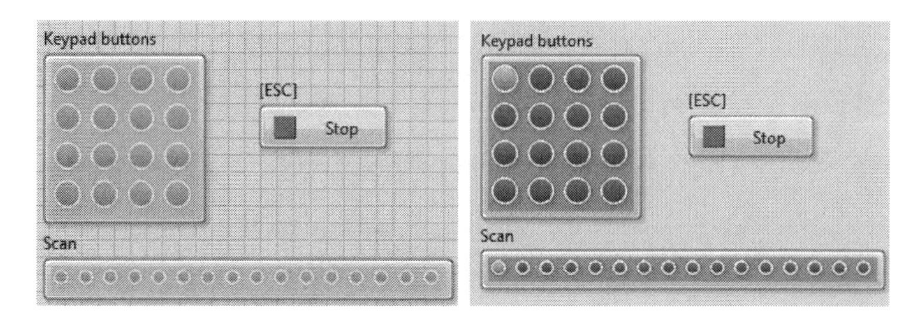

FIGURE 5.24 Final view of Keypad Front Panel.

5.2 Push Button Switch Interface

Push Button Switch—also called simple contact switches—serve as a basic user-interface to control the process. Figure 5.25 shows the Push Button Interface to myRIO wherein one end is connected to Pin 11(B/DIO0) and other end is connected to Pin 12(B/GND) with the help of a breadboard, 2 Jumper wires (M-F) and 2 wires.

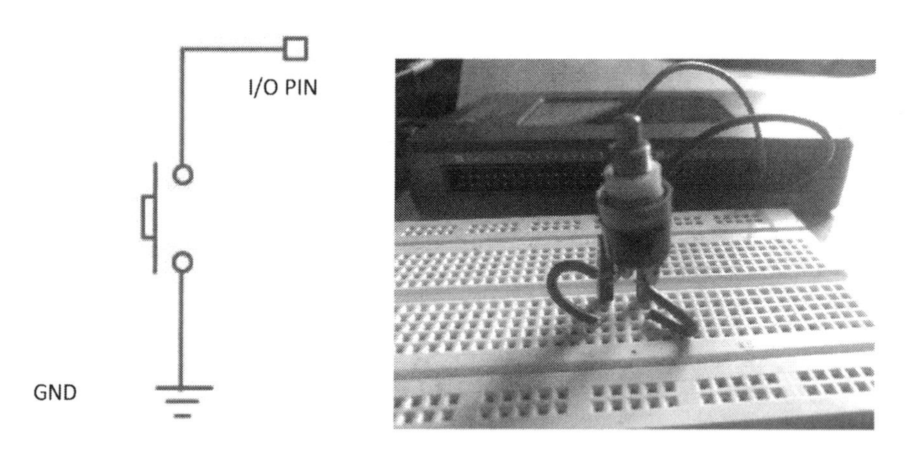

FIGURE 5.25 Push Button Switch Interface.

Implementation Steps

Step 1: In the Block Diagram window, right click on the screen. Under the Functions dialog box, click on Programming and then select Structures followed by the While Loop as shown in Figure 5.3. Now click and drag from the left top corner to the bottom right corner to form a rectangle.

Step 2: In the Block Diagram window, right click on the screen. Under the Functions dialog box, click on Programming and then select Structures followed by the Feedback Node as shown in Figure 5.26. It will look like [icon].

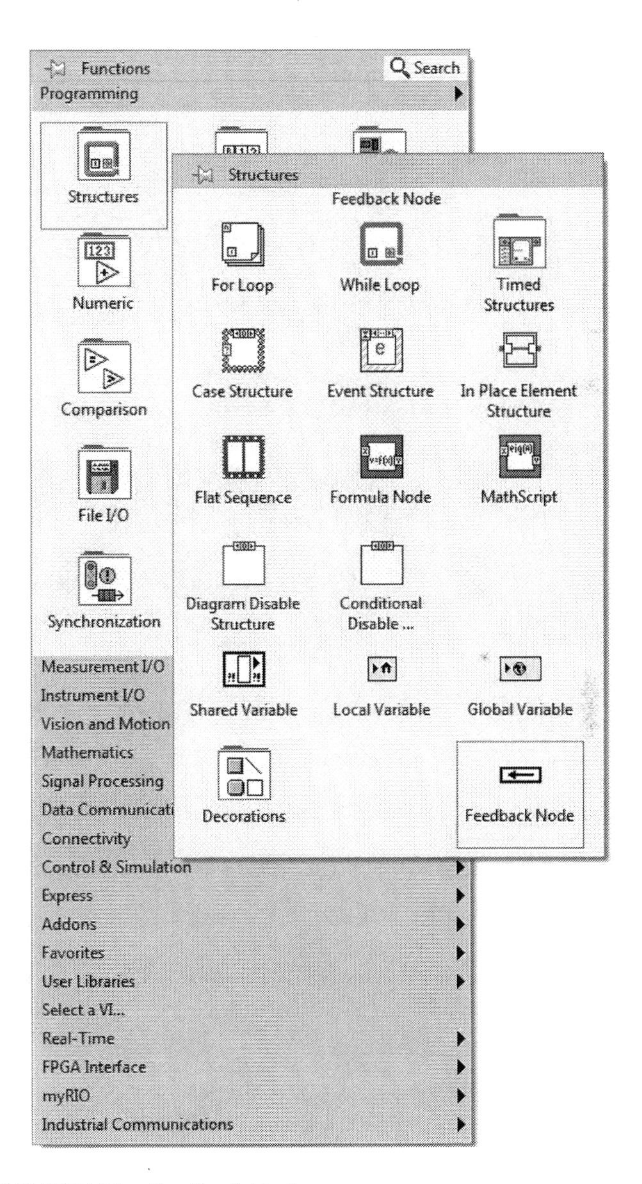

FIGURE 5.26 Feedback Node.

Step 3: Right click on the Feedback Node and select Properties. It will look like as shown in Figure 5.27.

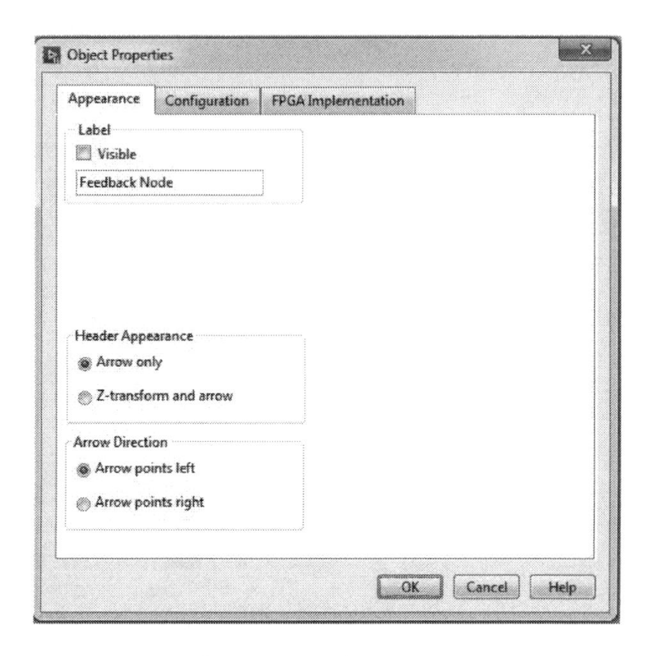

FIGURE 5.27 Feedback Node Object Properties.

Step 4: Under the Header Appearance of one of the feedback nodes, choose Z-transform and Arrow as well as under the Arrow Direction, choose Arrow points right as shown in Figure 5.28.

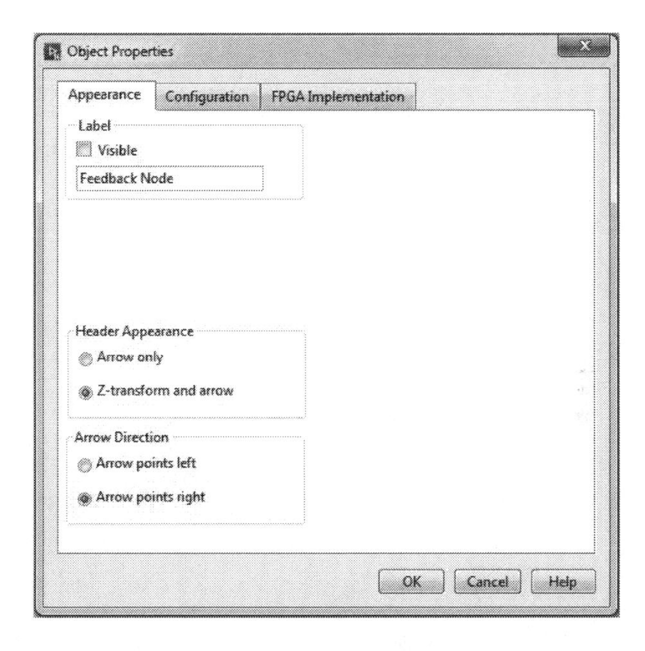

FIGURE 5.28 First Feedback Node Object Properties.

Step 5: Keep the Header Appearance of other feedback node as Arrow only and change the Arrow Direction to Arrow points right as shown in Figure 5.29.

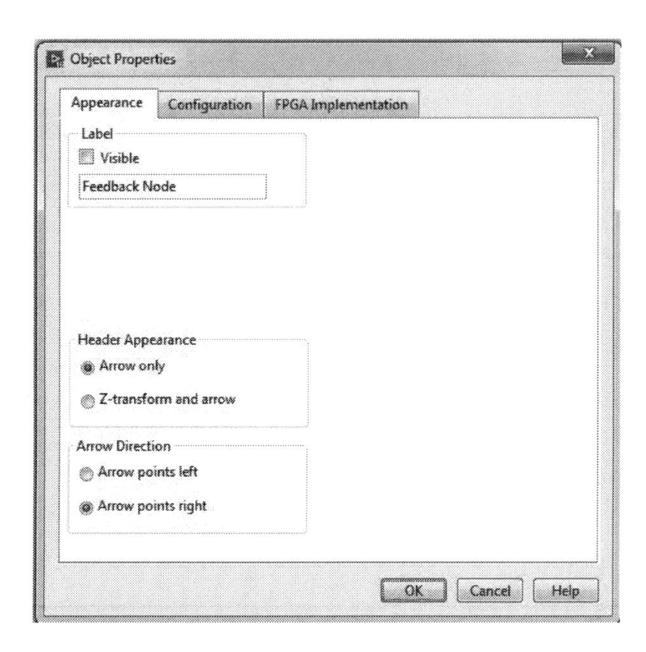

FIGURE 5.29 Second Feedback Node Object Properties.

Step 6: In the Block Diagram window, right click on the screen. Under the Functions dialog box, click on Programming, select Numeric and then Increment as shown in Figure 5.30.

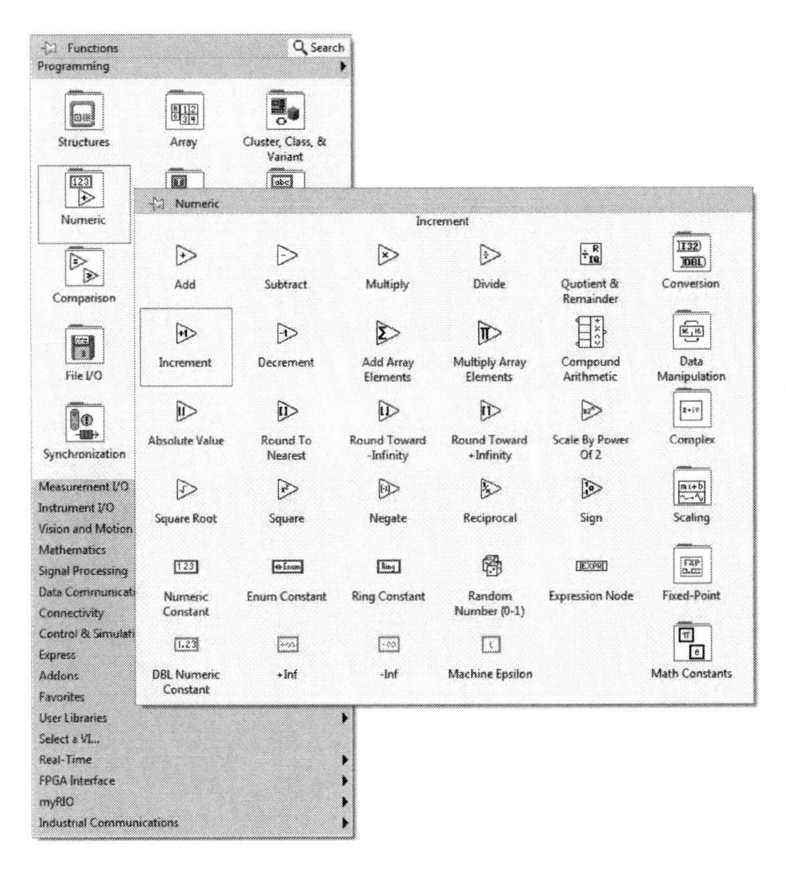

FIGURE 5.30 Increment.

Step 7: In the Block Diagram window, right click on the screen. Under the Functions dialog box, click on Programming, select Numeric and then Numeric Constant as shown in Figure 5.7. Add one more Numeric Constant. Change one of the values to 10.

Step 8: In the Block Diagram window, right click on the screen. Under the Functions dialog box, click on Programming and then select Boolean. Now click on And as shown in Figure 5.31.

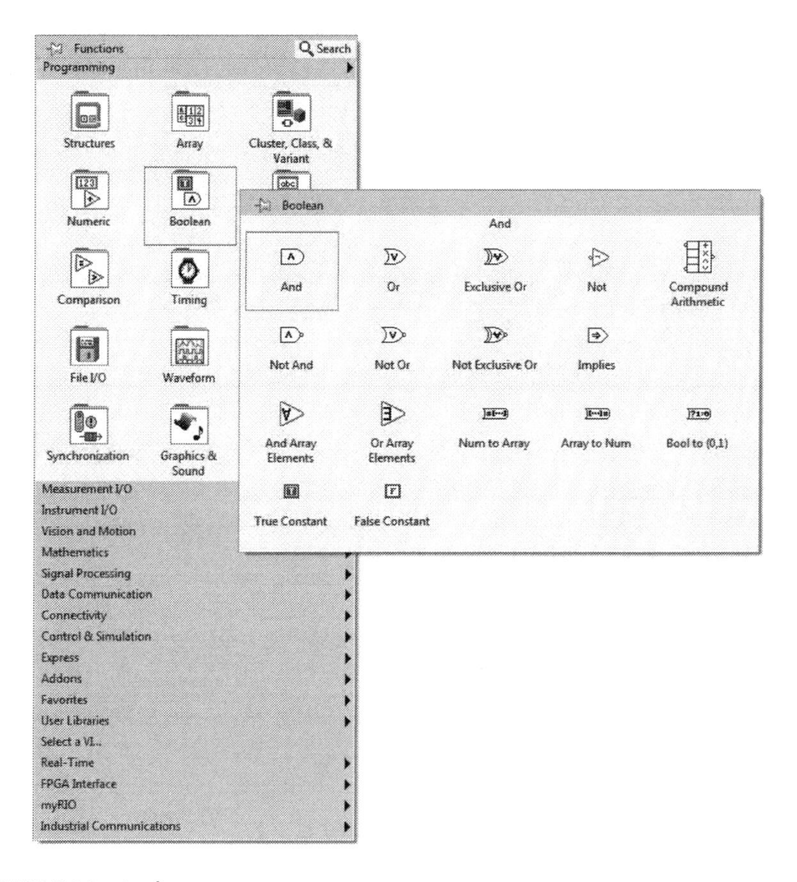

FIGURE 5.31 And.

Step 9: In the Block Diagram window, right click on the screen. Under the Functions dialog box, click on Programming and then select Boolean. Now click on Or as shown in Figure 5.8.

Step 10: In the Block Diagram window, right click on the screen. Under the Functions dialog box, click on Programming and then select Boolean. Now click on Not as shown in Figure 5.9. Add one more Not.

Step 11: In the Block Diagram window, right click on the screen. Under the Functions dialog box, click on Programming and then select Comparison. Now click on Select as shown in Figure 5.32 and place it in the block diagram.

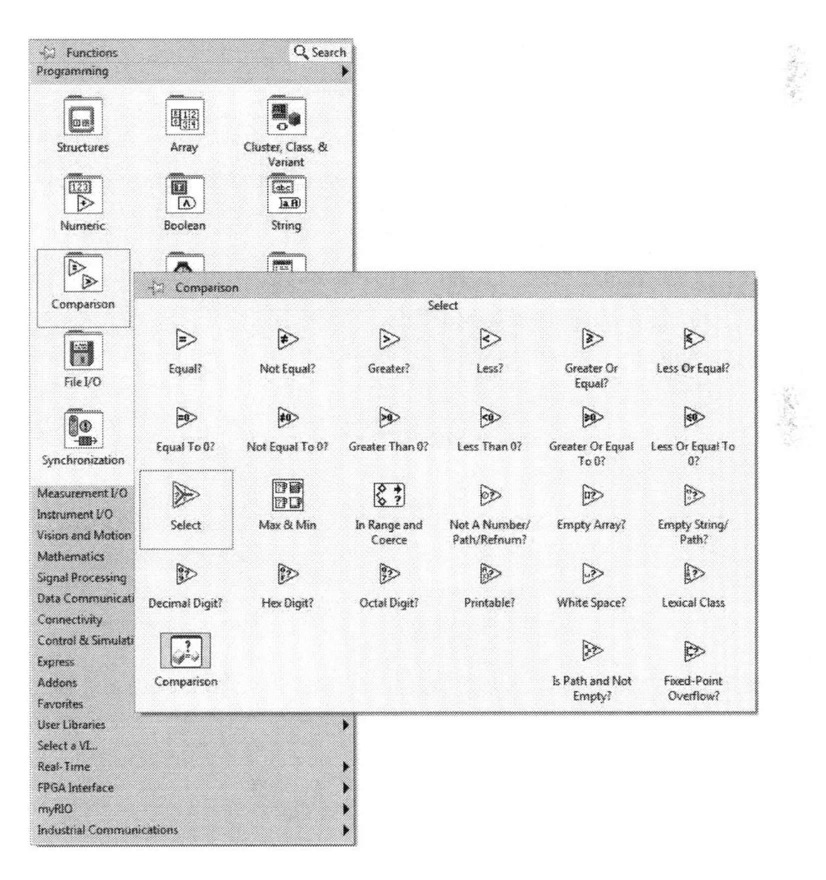

FIGURE 5.32 Select Comparison.

Step 12: In the Block Diagram window, right click on the screen. Under the Functions dialog box, click on Programming and then select Timing. Now click on Wait (ms) as shown in Figure 5.11.

Step 13: In the Block Diagram window, right click on the screen. Under the Functions dialog box, click on Programming and then select Dialog & User Interface. Now click on Simple Error Handler.vi as shown in Figure 5.12.

Step 14: In the Block Diagram window, right click on the screen. Under the Functions dialog box, click on myRIO and then select Digital In as shown in Figure 5.14.

Step 15: A Configure Digital Input (Default Personality) dialog box appears as shown in Figure 5.15. Change the channel to A/DIO0 (Pin 11). Click on the + button and change the channel to B/DIO0 (Pin 11). Click on the + button and change the channel to C/DIO0 and click on OK. Place the Digital In in the Block Diagram window. Now extend the Digital In downwards to display error out. Right click on error in (no error) and select Remove Input.

Step 16: In the Block Diagram window, right click on the screen. Under the Functions dialog box, click on myRIO and then Device Management. Now select Reset as shown in Figure 5.19.

Step 17: In the front panel window, right click on the screen. Under the Controls dialog box, click on Silver, Numeric and then select Numeric Indicator (Silver) as shown in Figure 5.33. Rename the label B/DIO0 presses. Right click on the Numeric Indicator (Silver) and click on Representation. Change the data type from Double Precision (DBL) to Long (I32) .

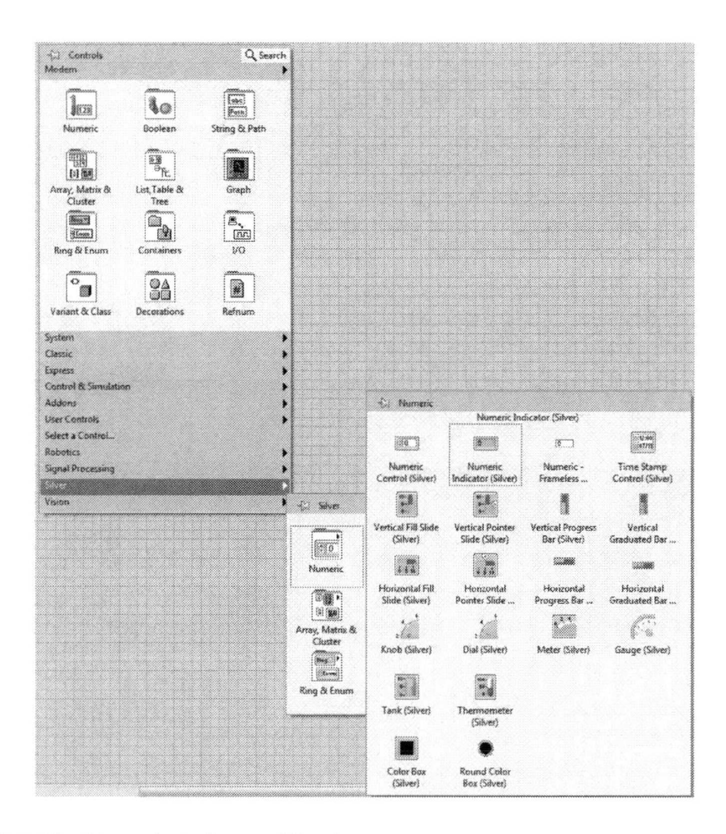

FIGURE 5.33 Numeric Indicator (Silver).

Step 18: In the Front Panel window, right click on the screen. Under the Controls dialog box, click on Silver and then Boolean. Now select LED (Silver) as shown in Figure 5.20.

Step 19: Right click on the LED and select properties. Rename the label A/DIO0 state. Select Show Boolean Text. Select Lock Text in Center. Rename On text as H and Off text as L. One can also change the height and width as shown in Figure 5.34.

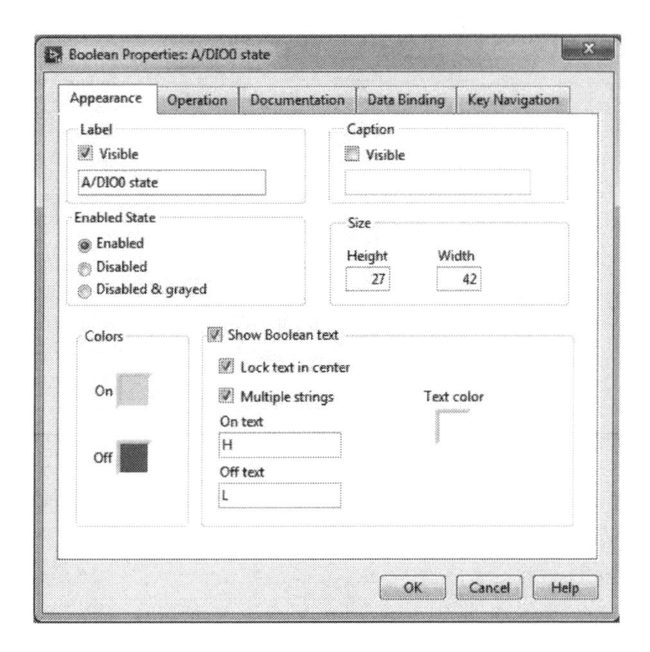

FIGURE 5.34 LED (Silver) Properties.

Step 20: In the Front Panel window, right click on the screen. Under the Controls dialog box, click on Silver and then Boolean. Now select Stop Button (Silver) as shown in Figure 5.21. Rename the label [ESC].

Step 21: Connect all the blocks as shown in the final view of the block diagram window in Figure 5.35.

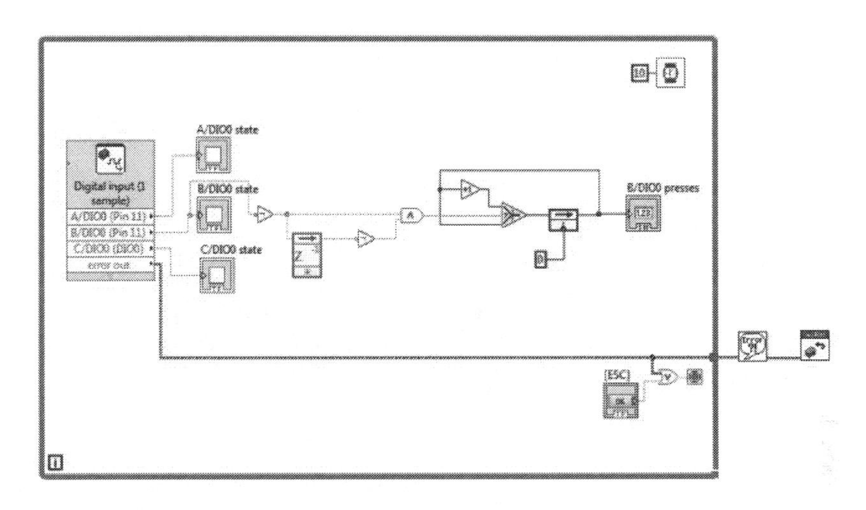

FIGURE 5.35 Final view of Push Button Block Diagram.

Step 22: The final view of the front panel window is shown in Figure 5.36.

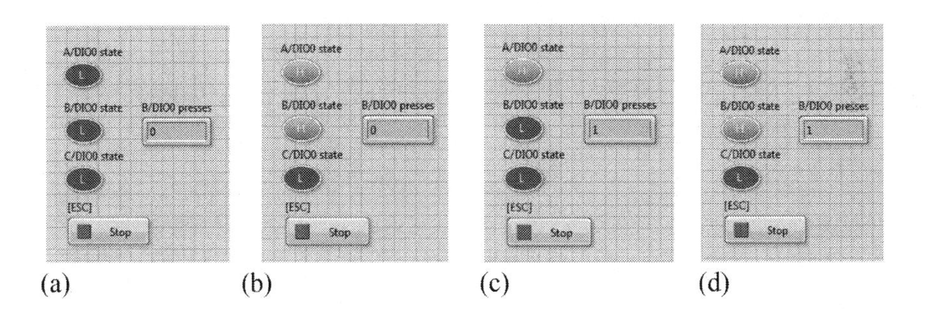

FIGURE 5.36 Final view of the different stages of the Push Button Front Panel. (a) Initial Stage. (b) After Running. (c) After pressing the Pushbutton. (d) After releasing the Pushbutton.

5.3 LED Interfacing

LEDs, or Light Emitting Diodes, provide simple yet essential visual indicators for system status and to indicate the error conditions. Figure 5.37 shows the LED interface to NI myRIO Starter Kit wherein Pin 33 (B/+3.3 V) of myRIO is connected to the anode of the LED with the help of a Jumper wire (M-F). The cathode of the LED is connected to one end of a 220 Ω resistor on a breadboard and the other end is connected to Pin 11 (B/DIO0) of myRIO with the help of another Jumper wire (M-F) [3].

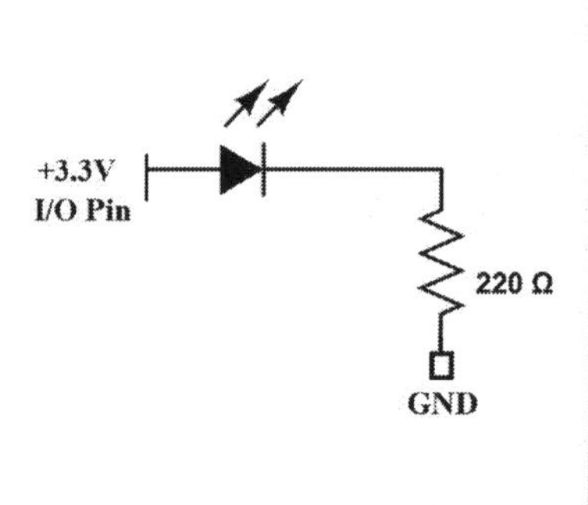

FIGURE 5.37 LED Interfacing.

Step 1: In the Block Diagram window, right click on the screen. Under the Functions dialog box, click on Programming and then select Structures followed by the While Loop as shown in Figure 5.3.

Step 2: Now click and drag from the left top corner to the bottom right corner to form a rectangle. Right click on the right side of the While Loop and select Add Shift Register.

Step 3: In the Block Diagram window, right click on the screen. Under the Functions dialog box, click on Programming and then select Numeric and then Numeric Constant as shown in Figure 5.7. Add two numeric constants. Change one of the values to 500 and the other to 10.

Step 4: In the Block Diagram window, right click on the screen. Under the Functions dialog box, click on Programming and then select Boolean. Now click on Or as shown in Figure 5.8.

Step 5: In the Block Diagram window, right click on the screen. Under the Functions dialog box, click on Programming and then select Boolean. Now click on Not as shown in Figure 5.9.

Step 6: In the Block Diagram window, right click on the screen. Under the Functions dialog box, click on Programming and then select Comparison. Now click on Select as shown in Figure 5.32 and place it in the block diagram.

Step 7: Repeat Step 6 and place another Select in the block diagram.

Step 8: In the Block Diagram window, right click on the screen. Under the Functions dialog box, click on Programming and then select Timing. Now click on Wait (ms) as shown in Figure 5.11.

Step 9: In the Block Diagram window, right click on the screen. Under the Functions dialog box, click on Programming and then select Dialog & User Interface. Now click on Simple Error Handler.vi as shown in Figure 5.12.

Step 10: In the Block Diagram window, right click on the screen. Under the Functions dialog box, click on myRIO and then select Digital Out as shown in Figure 5.38.

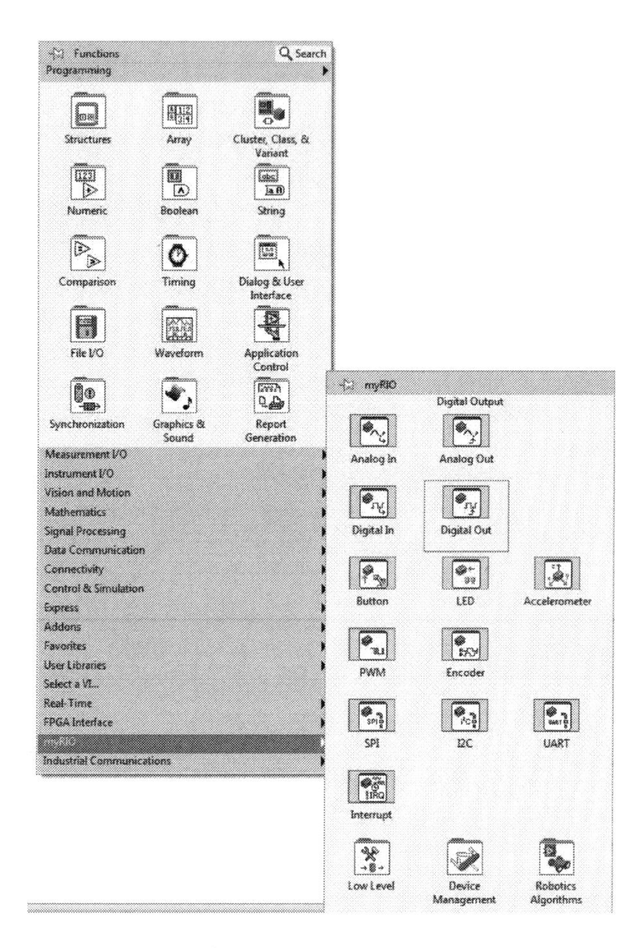

FIGURE 5.38 Digital Out.

Step 11: A Configure Digital Output (Default Personality) Dialog Box appears as shown in Figure 5.39. Change the channel to B/DIO0 (Pin 11) and click on OK. Place the Digital Out in the Block Diagram window.

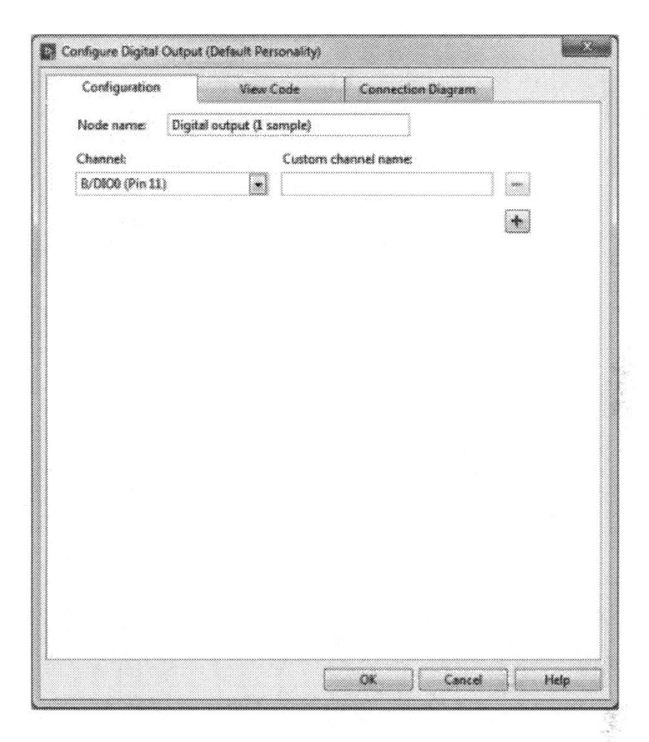

FIGURE 5.39 Configure Digital Output (Default Personality) dialog box.

Step 12: In the Block Diagram window, right click on the screen. Under the Functions dialog box, click on myRIO and then Device Management. Now select Reset as shown in Figure 5.19.

Step 13: In the Front Panel window, right click on the screen. Under the Controls dialog box, click on Silver and then Boolean. Now select Push Button (Silver) as shown in Figure 5.40 and place it in the front panel. Rename the label Enable Blinker.

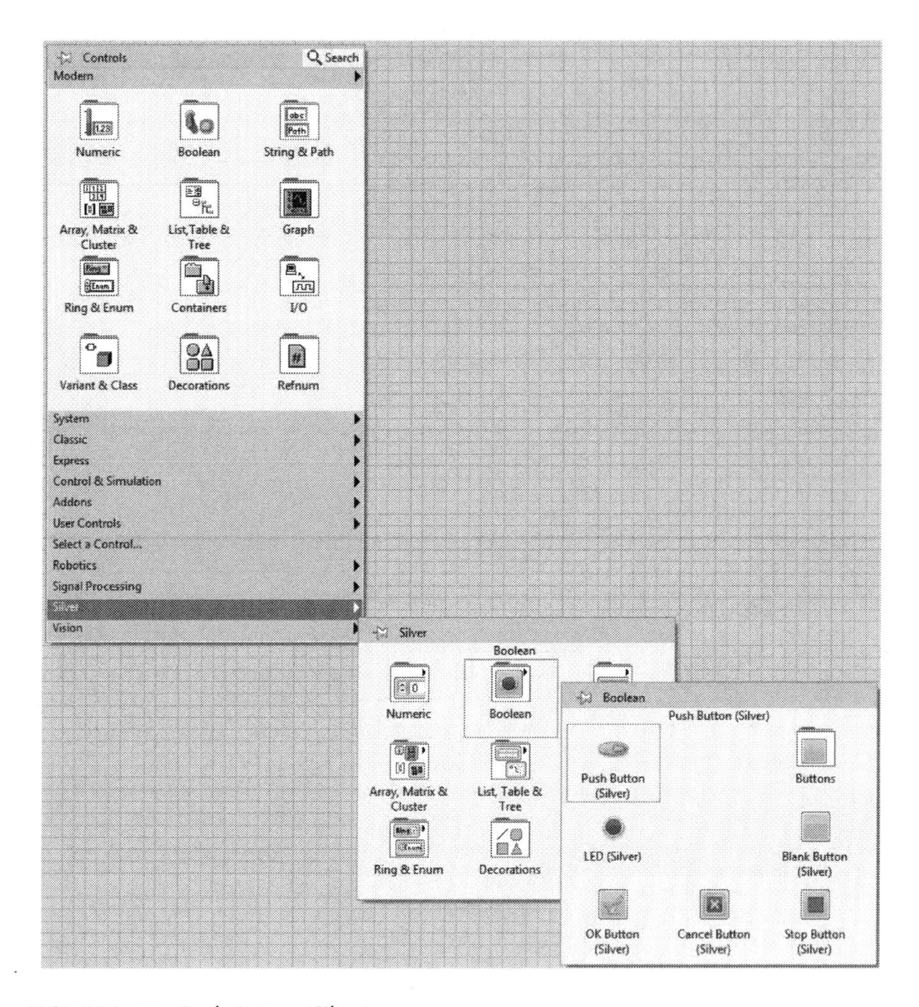

FIGURE 5.40 Push Button (Silver).

Step 14: Repeat Step 13 and place another Push Button (Silver) in the front panel. Rename the label Digital Level.

Step 15: In the Front Panel window, right click on the screen. Under the Controls dialog box, click on Silver and then Boolean. Now select LED (Silver) as shown in Figure 5.20. Rename the label Digital Output State.

Step 16: In the Front Panel window, right click on the screen. Under the Controls dialog box, click on Silver and then Boolean. Now select Stop Button (Silver) as shown in Figure 5.21. Rename the label [ESC].

Step 17: Connect all the blocks as shown in the final view of the block diagram window in Figure 5.41.

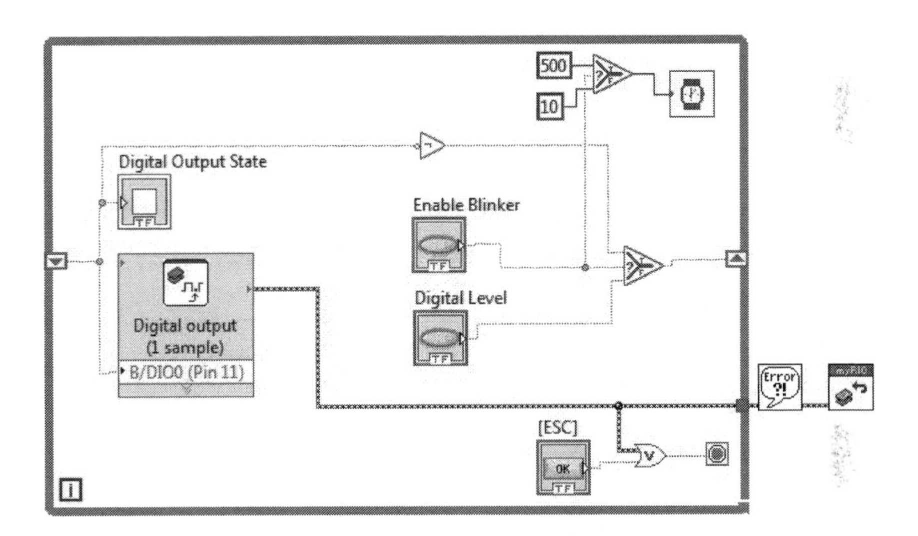

FIGURE 5.41 Final view of LED Interfacing Block Diagram.

Step 18: The final view of the front panel window is shown in Figure 5.42.

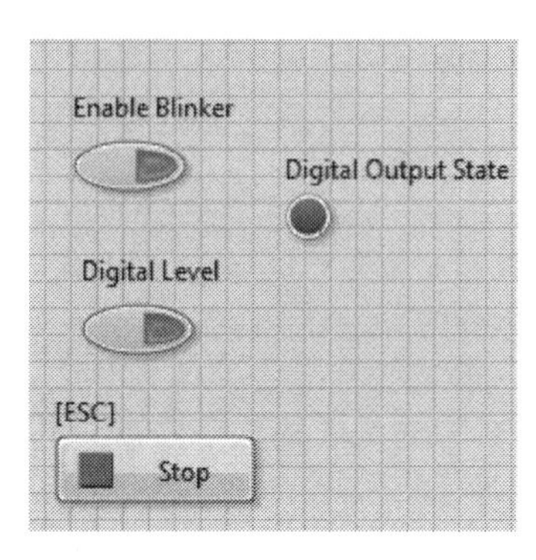

FIGURE 5.42 Final view of LED interface Front Panel.

5.4 Seven-segment LED Display

The *7-segment display* consists of seven LEDs (hence its name) arranged properly in a rectangular fashion. Each of the seven LEDs is called a segment because when lighted up, the segment forms the numerical digit. The interface of 7-segment to myRIO kit is shown in Figure 5.43. Here we have an interface of four segments which are then multiplexed so as to reduce the number of I/O lines. Here we require only 12 I/O lines (8 data and 4 control or segment selections).

In order to interface the 7-segment Display to myRIO, 9 Jumper Wires (M-F) are connected with the help of a breadboard in the following way:

A – pin 11 (B/DIO0)
B – pin 13 (B/DIO1)
C – pin 15 (B/DIO2)
D – pin 17 (B/DIO3)
E – pin 19 (B/DIO4)
F – pin 21 (B/DIO5)
G – pin 23 (B/DIO6)
decimal – pin 25 (B/DIO7)
D1 – pin 33 (B/+3.3V)

wherein A, B, C, D, E, F, and G are the 7 segments, decimal is the decimal point, and D1 is the first digit selected. If the other Digits D2, D3, or D4 are to be selected, connect them using wires to D1 on the breadboard, so that it is finally connected to B/+3.3V of myRIO.

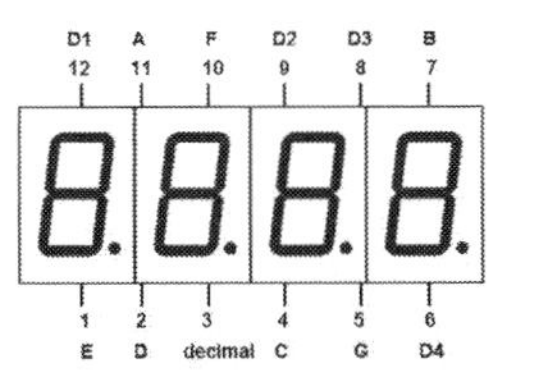

FIGURE 5.43 Seven-segment LED display Interfacing.

Implementation Steps

Step 1: In the Block Diagram window, right click on the screen. Under the Functions dialog box, click on Programming and then select Structures followed by the While Loop as shown in Figure 5.3. Now click and drag from the left top corner to the bottom right corner to form a rectangle.

Step 2: In the Block Diagram window, right click on the screen. Under the Functions dialog box, click on Programming and then select Array followed by Index Array as shown in Figure 5.44.

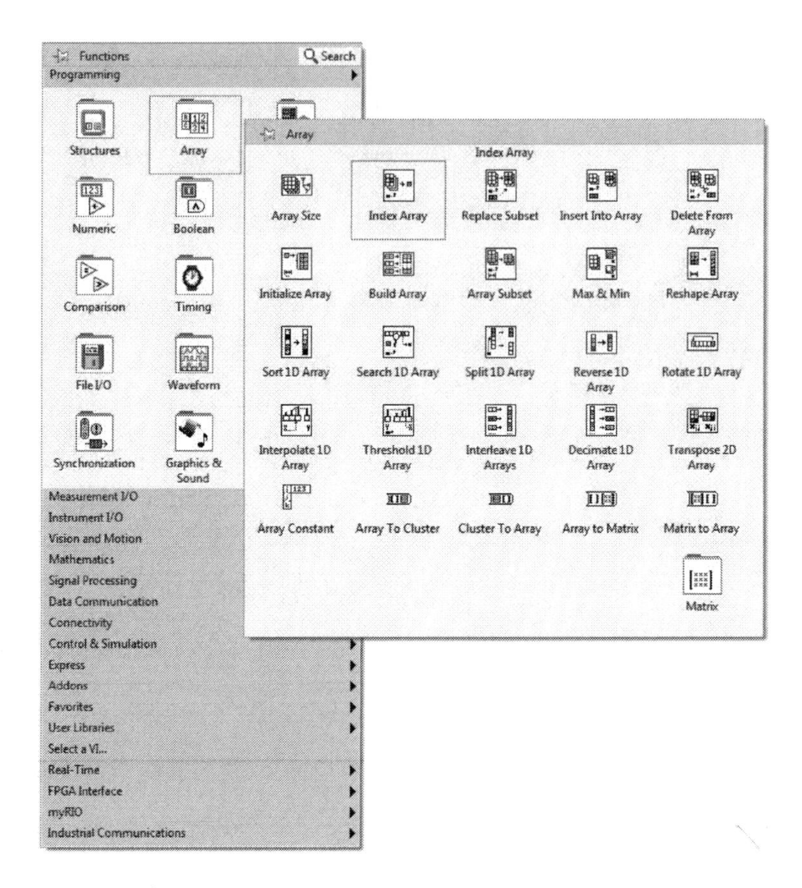

FIGURE 5.44 Index Array.

Step 3: After placing it in the block diagram, extend the Index Array by dragging the bottom edge in the bottom direction such that 8 indices are visible.

Step 4: In the Block Diagram window, right click on the screen. Under the Functions dialog box, click on Programming, select Numeric and then Numeric Constant as shown in Figure 5.7. Change the value to 100.

Step 5: In the Block Diagram window, right click on the screen. Under the Functions dialog box, click on Programming and then select Boolean. Now click on Or as shown in Figure 5.8.

Step 6: In the Block Diagram window, right click on the screen. Under the Functions dialog box, click on Programming and then select Boolean. Now click on Not as shown in Figure 5.9.

Step 7: In the Block Diagram window, right click on the screen. Under the Functions dialog box, click on Programming and then select Timing. Now click on Wait (ms) as shown in Figure 5.11.

Step 8: In the Block Diagram window, right click on the screen. Under the Functions dialog box, click on Programming and then select Dialog & User Interface. Now click on Simple Error Handler.vi as shown in Figure 5.12.

Step 9: In the Block Diagram window, right click on the screen. Under the Functions dialog box, click on myRIO and then select Digital Out as shown in Figure 5.38.

Step 10: A Configure Digital Output (Default Personality) dialog box appears as shown in Figure 5.39. Change the channel to B/DIO0 (Pin 11) and click on the + button. Add another seven more channels from B/DIO1 (Pin 13) upto B/DIO7 (Pin 25) and click on OK. Place the Digital Out in the Block Diagram window. Now extend the Digital Out downwards to display error out. Right click on error in and select Remove Input.

Step 11: In the Block Diagram window, right click on the screen. Under the Functions dialog box, click on myRIO and then Device Management. Now select Reset as shown in Figure 5.19.

Step 12: In the Front Panel window, right click on the screen. Under the Controls dialog box, click on Silver and then Boolean. Now select Push Button (Silver) as shown in Figure 5.40 and place it in the front panel.

Step 13: Repeat Step 12 and place another seven more Push Button (Silver) in the front panel.

Step 14: In the Front Panel window, right click on the screen. Under the Controls dialog box, click on Silver and then Boolean. Now select Stop Button (Silver) as shown in Figure 5.21. Now right click on the Stop button and select Properties; change the label to [ESC].

Step 15: Connect all the blocks as shown in the final view of the Block Diagram window in Figure 5.45.

Step 16: The final view of the Front Panel window is shown in Figure 5.46 to display “1.”

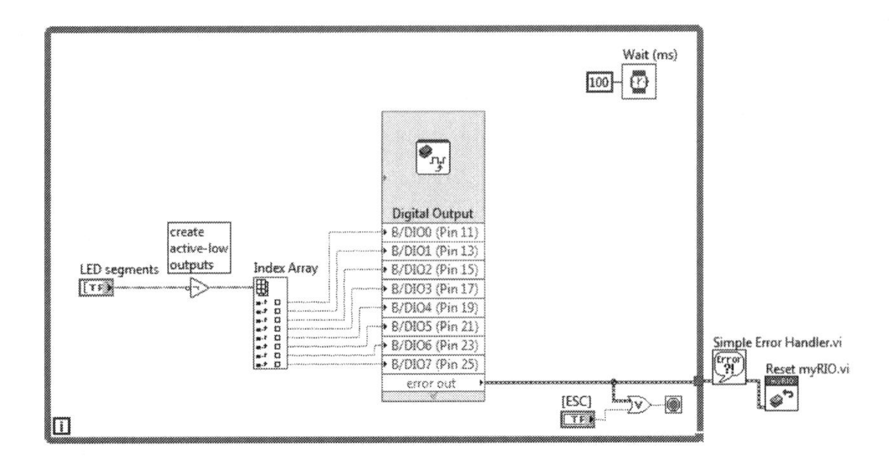

FIGURE 5.45 Final view of 7 segment LED display Block Diagram.

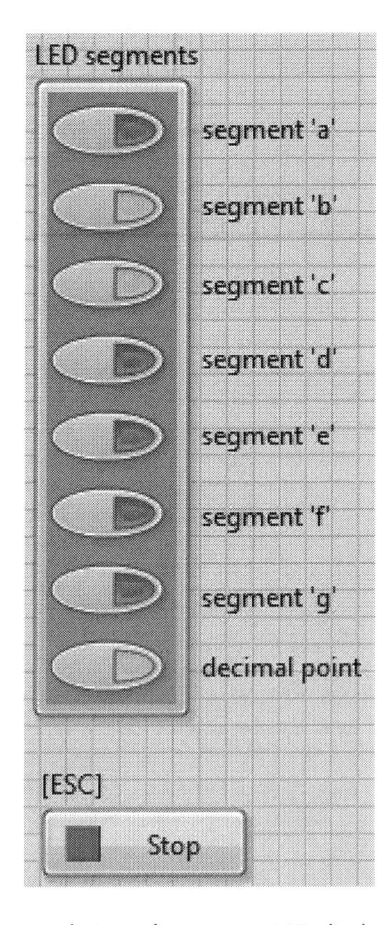

FIGURE 5.46 Final view of 7 segment LED display Front Panel.

5.5 UART Interface

There are various serial communication interfaces available.

A Universal Asynchronous Receiver/Transmitter (UART) is a circuitry responsible for implementing serial communication. Essentially, the UART acts as an interface between parallel and serial interfaces. UART has two main lines—Receive (RX) and Transmit (TX). Here data can be transferred at different baud rates. Use a Digilent PmodCLS LCD to perform this experiment along with 3 Jumper Wires (F-F) and 4 Jumpers. Connect the Jumpers as shown in Figure 5.47. Connect J2/V of the LCD to pin 33 (B/+3.3V) of myRIO using a Jumper Wire, connect J2/G of the LCD to pin 30 (B/GND) of myRIO using a Jumper Wire and connect J2/RX of the LCD to pin 14 (B/UART.TX) of myRIO using a Jumper Wire [2].

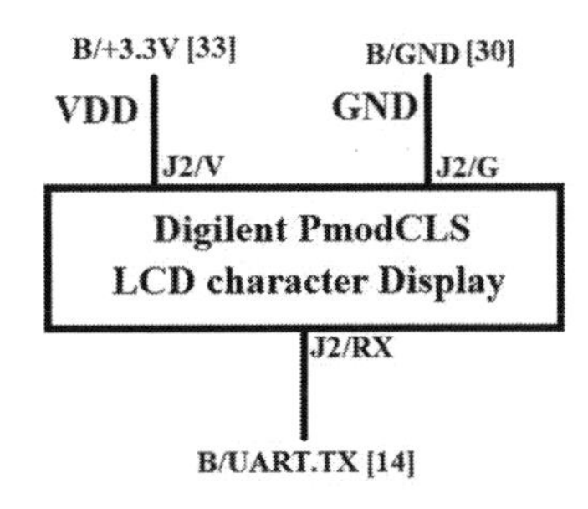

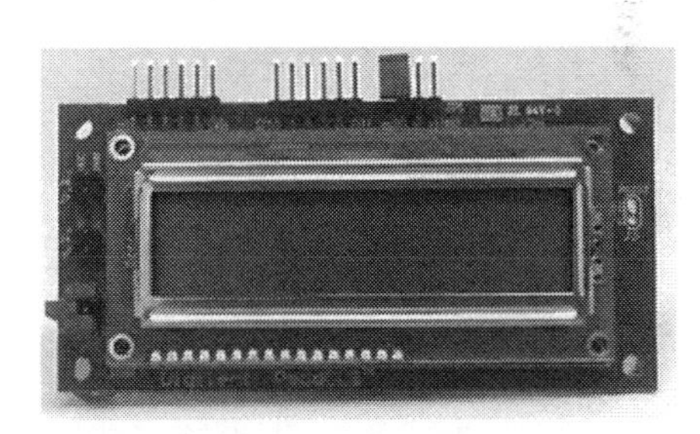

FIGURE 5.47 UART interface.

Implementation Steps

UART based LCD interface

Step 1: In the Block Diagram window, right click on the screen. Under the Functions dialog box, click on Programming and then select Structures followed by the While Loop as shown in Figure 5.3. Now click and drag from the left top corner to the bottom right corner to form a rectangle.

Step 2: In the Block Diagram window, right click on the screen. Under the Functions dialog box, click on Programming and then select Numeric and then Numeric Constant as shown in Figure 5.7. Change the value to 200.

Step 3: In the Block Diagram window, right click on the screen. Under the Functions dialog box, click on Programming and then select Boolean. Now click on Or as shown in Figure 5.8.

Step 4: In the Block Diagram window, right click on the screen. Under the Functions dialog box, click on Programming and then select Boolean. Now click on Bool to (0,1) as shown in Figure 5.48.

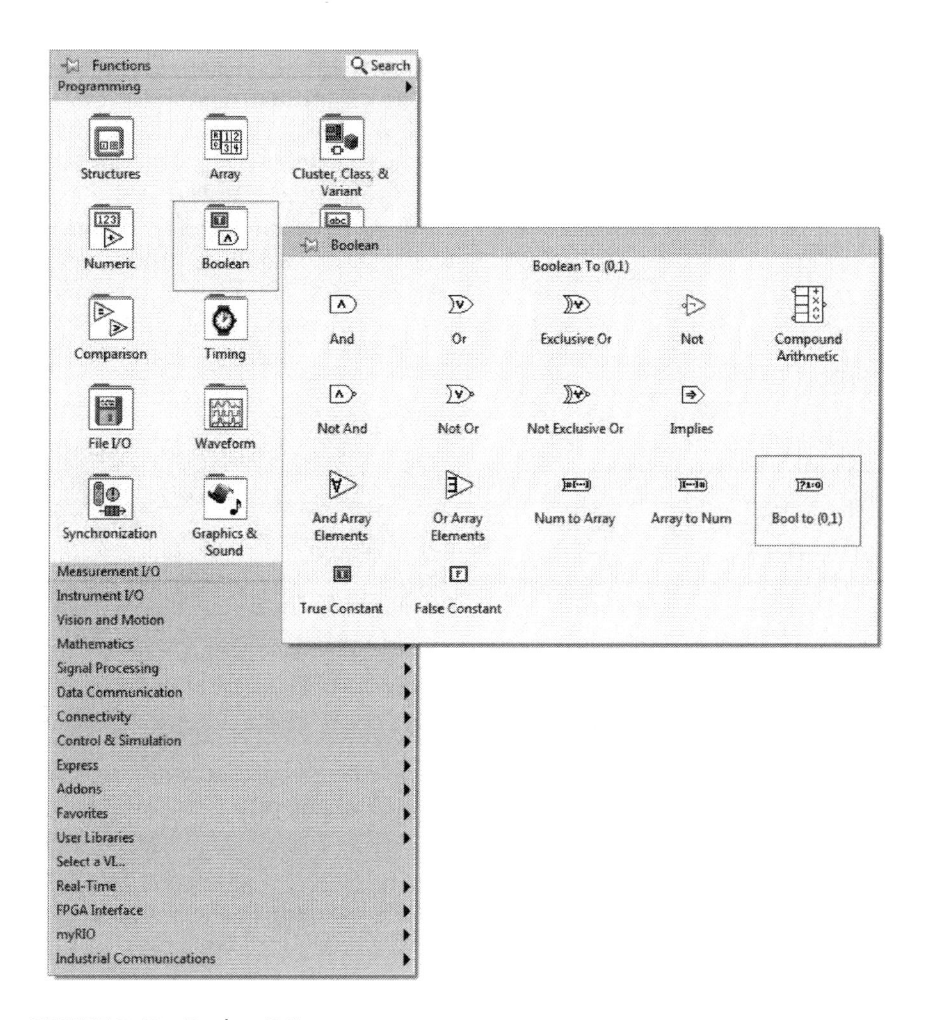

FIGURE 5.48 Bool to (0,1).

Step 5: In the Block Diagram window, right click on the screen. Under the Functions dialog box, click on Programming and then select String. Now click on Format Into String as shown in Figure 5.49.

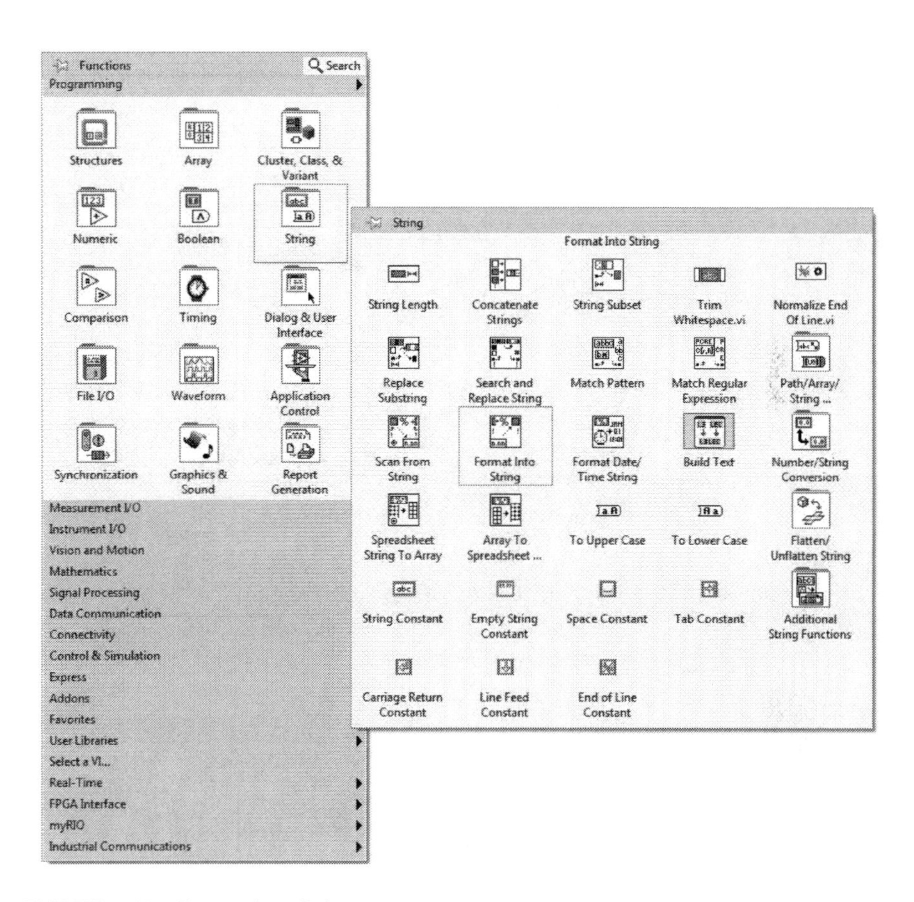

FIGURE 5.49 Format Into String.

Step 6: Right click on the Format Into String and click on Properties. The Edit Format String appears as below. Click on Add New Operation four times. Type in "X: %5.2f Y: %5.2f Z: %5.2f Button: %d" in the Corresponding format string box as shown in Figure 5.50. Click on OK.

It will look like .

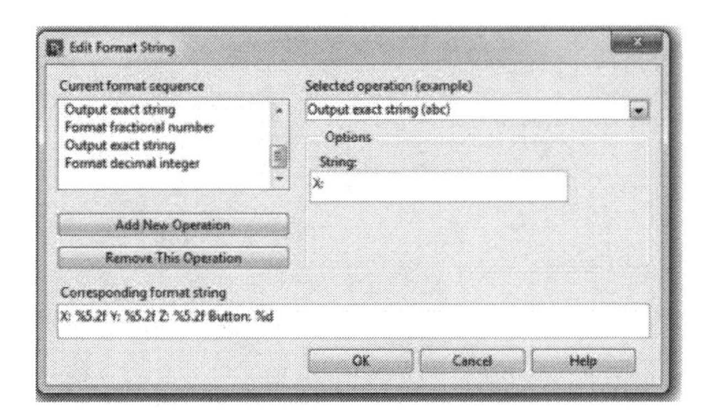

FIGURE 5.50 Edit Format String.

Step 7: In the Block Diagram window, right click on the screen. Under the Functions dialog box, click on Programming and then select Timing. Now click on Wait (ms) as shown in Figure 5.11.

Step 8: In the Block Diagram window, right click on the screen. Under the Functions dialog box, click on Programming and then select Dialog & User Interface. Now click on Simple Error Handler.vi as shown in Figure 5.12.

Step 9: In the Block Diagram window, right click on the screen. Under the Functions dialog box, click on myRIO and then select Button as shown in Figure 5.51.

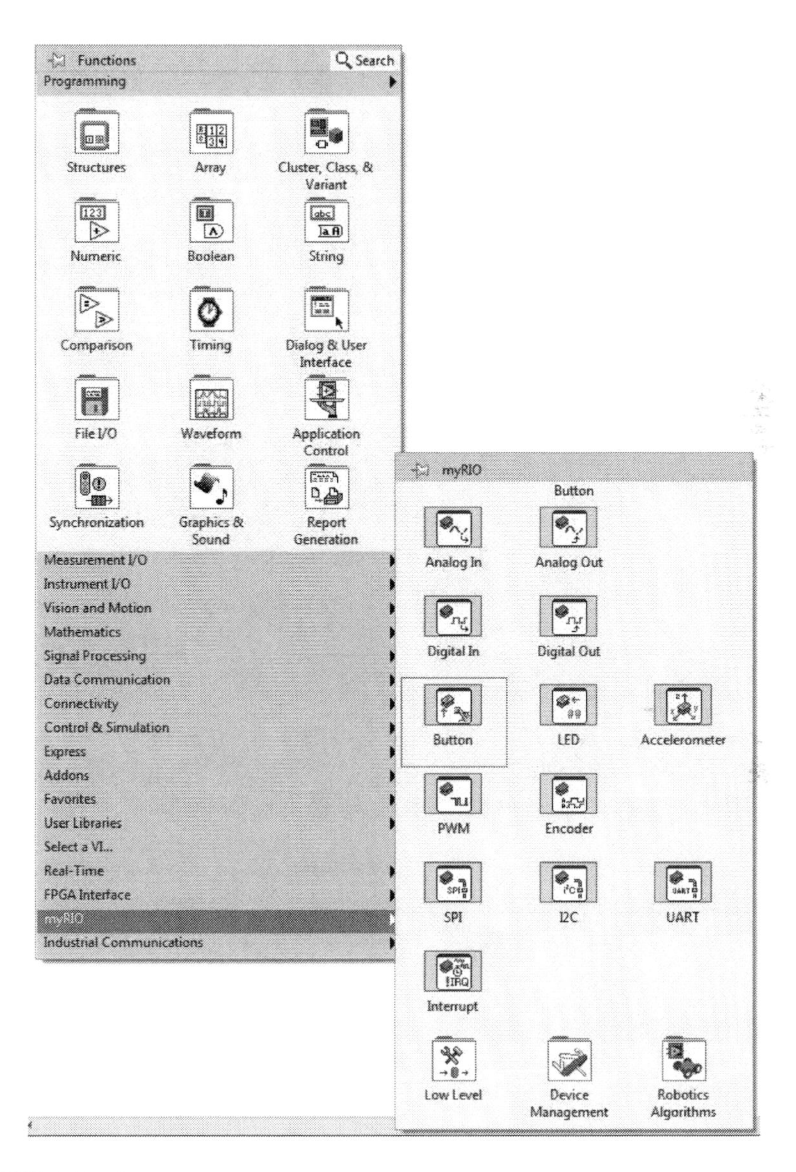

FIGURE 5.51 Button.

Step 10: A Configure Button (Default Personality) dialog box appears as shown in Figure 5.52. Click on OK. Now place it in the block diagram and extend the button by dragging the bottom edge downwards such that error in and error out appear. It will look like .

FIGURE 5.52 Configure Button (Default Personality).

Step 11: In the Block Diagram window, right click on the screen. Under the Functions dialog box, click on myRIO and then select Accelerometer as shown in Figure 5.53.

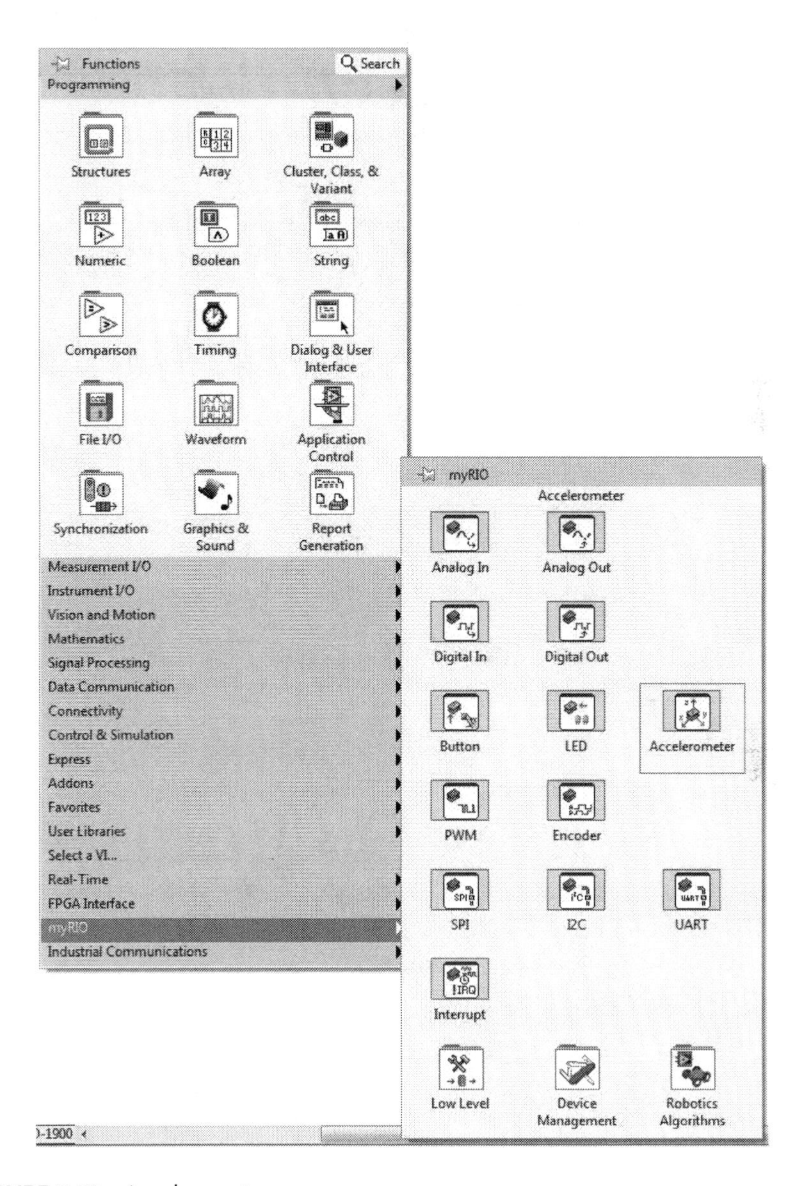

FIGURE 5.53 Accelerometer.

Step 12: A Configure Accelerometer (Default Personality) dialog box appears as shown in Figure 5.54. Click on OK. Now place it in the block diagram and extend the accelerometer by dragging the bottom edge downwards such that error in and error out appear. It will look like

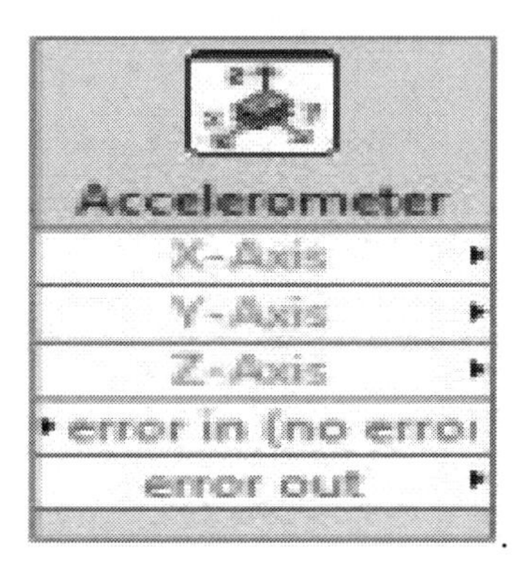

.

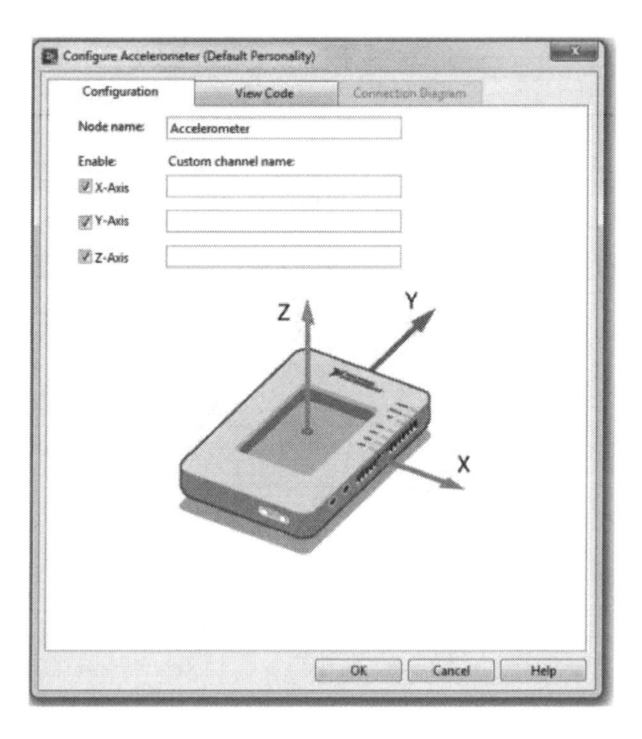

FIGURE 5.54 Configure Accelerometer (Default Personality).

Step 13: In the Block Diagram window, right click on the screen. Under the Functions dialog box, click on myRIO and then select UART as shown in Figure 5.55.

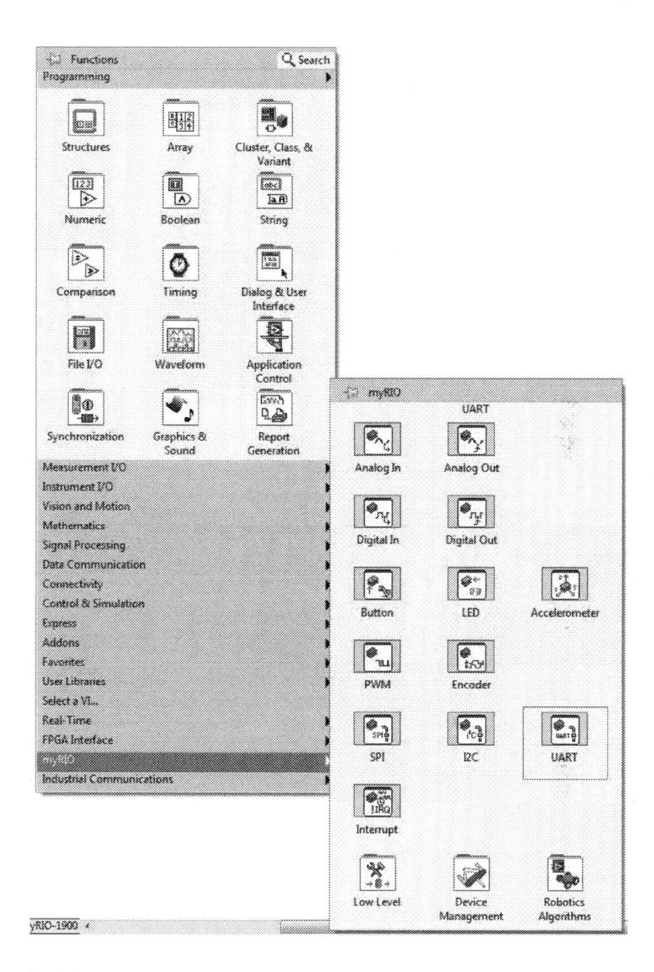

FIGURE 5.55 UART.

Step 14: A Configure UART (Default Personality) dialog box appears as shown in Figure 5.56. Click on OK. Now place it in the block diagram and extend the UART by dragging the bottom edge downwards such that error in and error out appear. It will look like 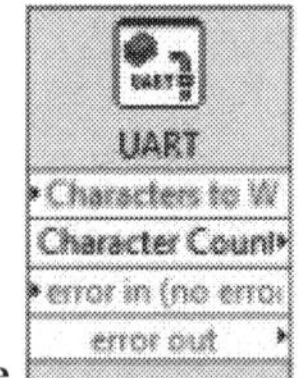

.

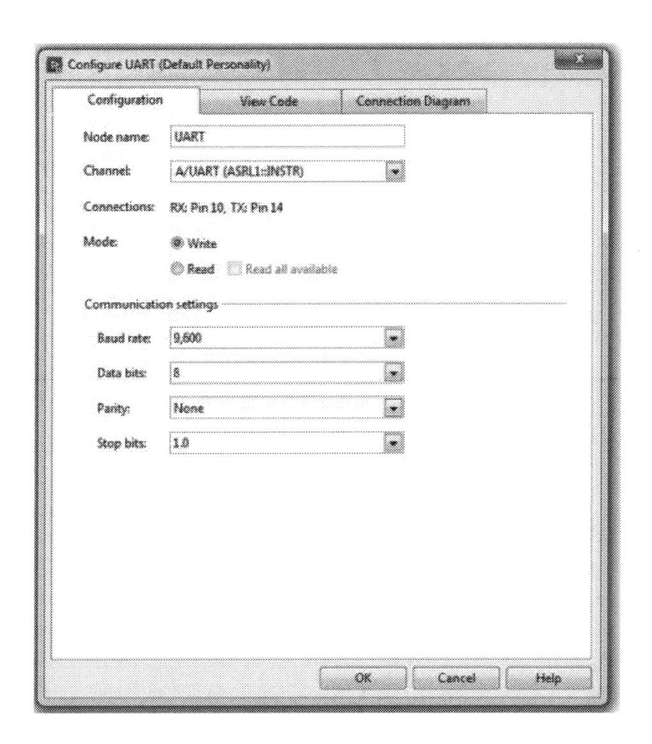

FIGURE 5.56 Configure UART (Default Personality).

Step 15: Repeat Step 14 and place another UART in the block diagram. Right click on the first UART at the error in (no errors) and select Remove Input. Right click on the input of the UART, Characters to Write and select Create and then Constant. Type in "[0h[j". Now right click on the constant and select "\" Codes Display. It will now look like "\1B[0h\1B[j".

Step 16: In the Block Diagram window, right click on the screen. Under the Functions dialog box, click on myRIO and then Device Management. Now select Reset as shown in Figure 5.19.

Step 17: In the Front Panel window, right click on the screen. Under the Controls dialog box, click on Silver and then Boolean. Now select Stop Button (Silver) as shown in Figure 5.21. Rename the label [ESC].

Step 18: In the Front Panel window, right click on the screen. Under the Controls dialog box, click on Silver and then String & Path. Now select String Indicator (Silver) as shown in Figure 5.57. Rename the label LCD display.

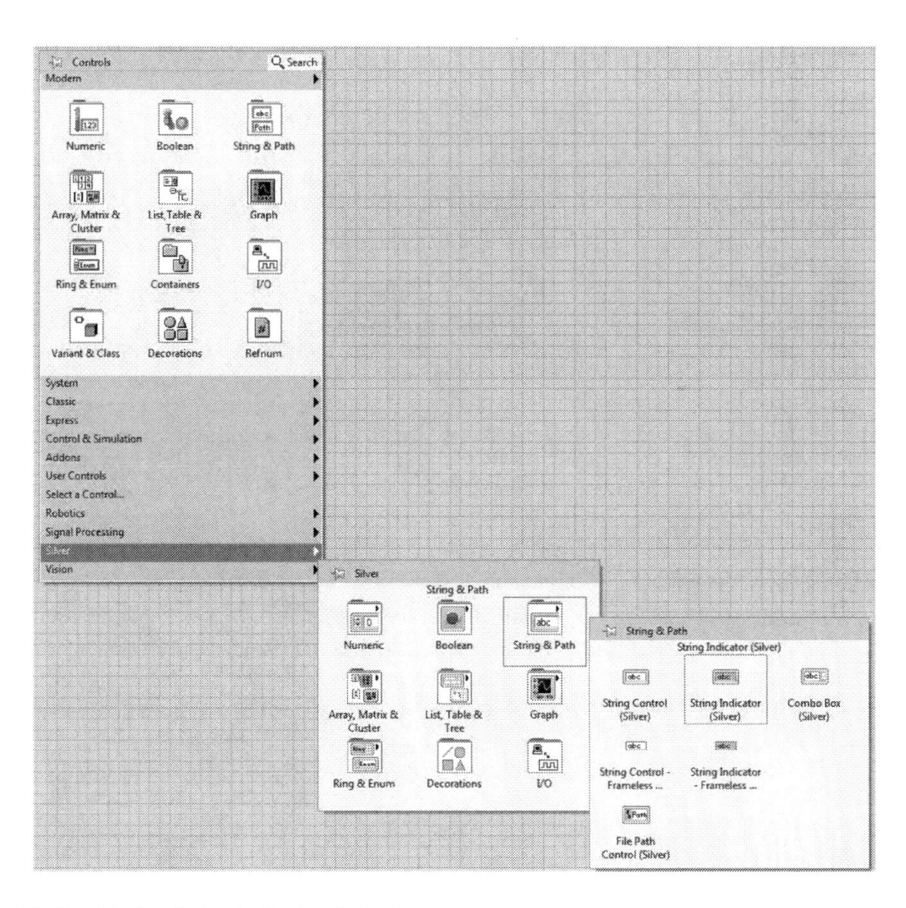

FIGURE 5.57 String Indicator (Silver).

Step 19: Connect all the blocks as shown in the final view of the block diagram window in Figure 5.58.

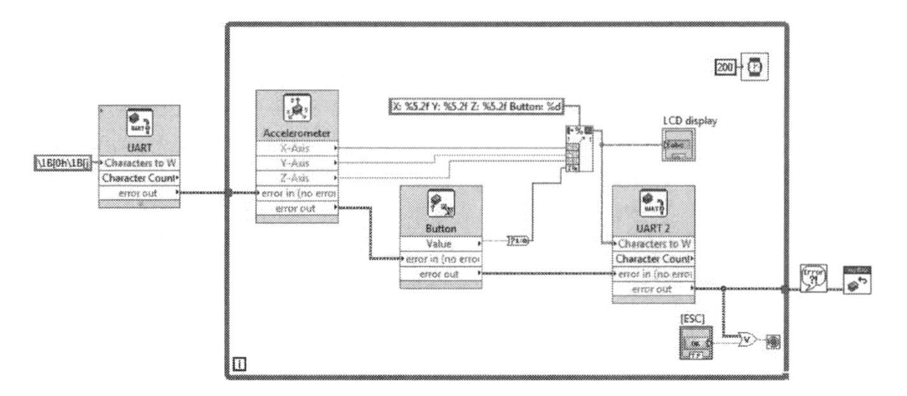

FIGURE 5.58 Final view of UART interface based LCD Block Diagram.

Step 20: The final view of the front panel window is shown in Figure 5.59.

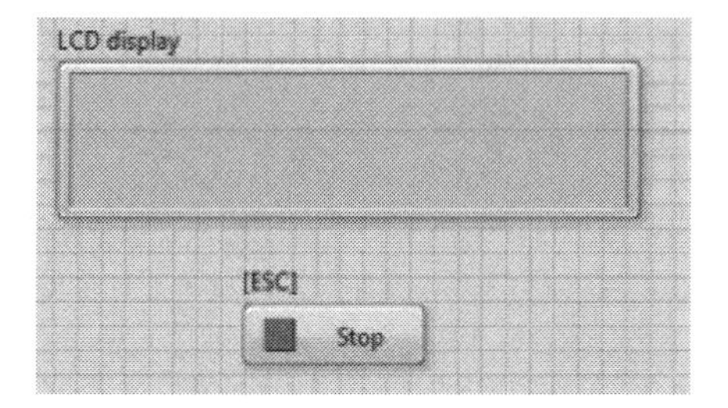

FIGURE 5.59 Final view of front panel UART based LCD.

5.6 I2C-bus Interface

I2C is a protocol which allows serial data transfer using two-wire interface to connect low-speed devices such as microcontrollers, EEPROMs, A/D and D/A converters, I/O interfaces and other similar peripherals in embedded systems. This protocol was designed by Philips. Use a Digilent PmodCLS LCD to perform this experiment along with 4 Jumper Wires (F-F) and 4 Jumpers. Connect the Jumpers as shown in Figure 5.60. Connect J2/V of the LCD to pin 33 (B/+3.3 V) of my RIO using a Jumper Wire, connect J2/G of the LCD to pin 30 (B/GND) of my RIO using a Jumper Wire, connect J2/SD of the LCD to pin 34 (B/I2C.SDA) of my RIO using a Jumper Wire and connect J2/SC of the LCD to pin 32 (B/I2C.SCL) of my RIO using a Jumper Wire.

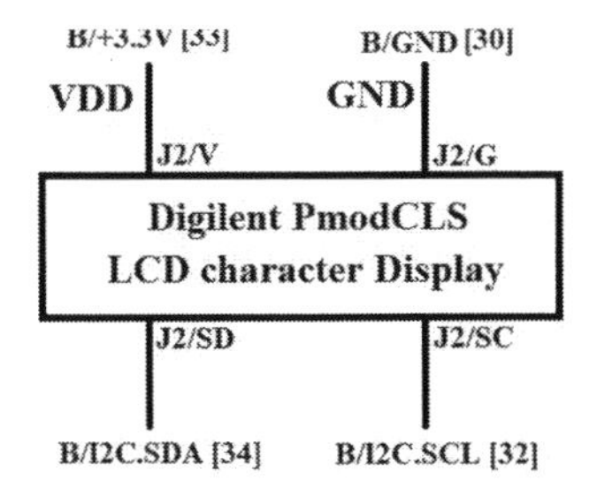

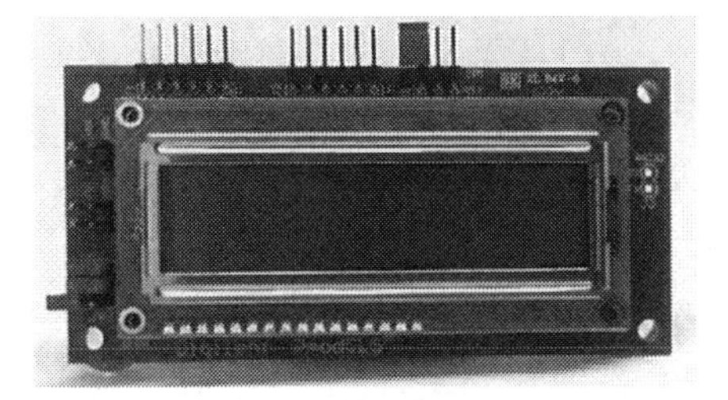

FIGURE 5.60 I2C interface.

Implementation Steps

I2C based LCD Interface

Step 1: In the Block Diagram window, right click on the screen. Under the Functions dialog box, click on Programming and then select Structures followed by the While Loop as shown in Figure 5.3. Now click and drag from the left top corner to the bottom right corner to form a rectangle.

Step 2: In the Block Diagram window, right click on the screen. Under the Functions dialog box, click on Programming and then select Numeric and then Conversion. Now select String to Byte Array as shown in Figure 5.61.

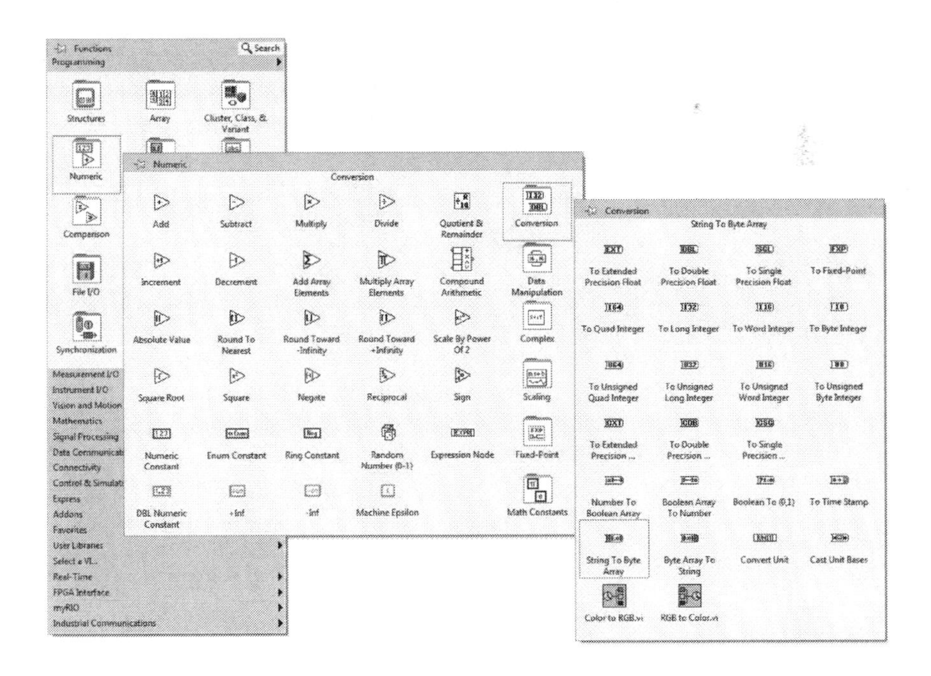

FIGURE 5.61 String to Byte Array.

Step 3: Repeat Step 2 and place another String to Byte Array in the Block Diagram. Right click on the String to Byte Array and select Create and then Constant. Type in "[0h[j". Now right click on the constant and select "\" Codes Display. It will now look like "\1B[0h\1B[j".

Step 4: In the Block Diagram window, right click on the screen. Under the Functions dialog box, click on Programming and then select Numeric and then Numeric Constant as shown in Figure 5.7. Change the value to 200.

Step 5: In the Block Diagram window, right click on the screen. Under the Functions dialog box, click on Programming and then select Boolean. Now click on Or as shown in Figure 5.8.

Step 6: In the Block Diagram window, right click on the screen. Under the Functions dialog box, click on Programming and then select Boolean. Now click on Bool to (0,1) as shown in Figure 5.48.

Step 7: In the Block Diagram window, right click on the screen. Under the Functions dialog box, click on Programming and then select String. Now click on Format Into String as shown in Figure 5.49.

Step 8: Right click on Format Into String and click on Properties. Edit Format String appears as below. Click on Add New Operation four times. Type in "X: %5.2f Y: %5.2f Z: %5.2f Button: %d" in the Corresponding format string box as shown in Figure 5.50. Click on OK. It will look like

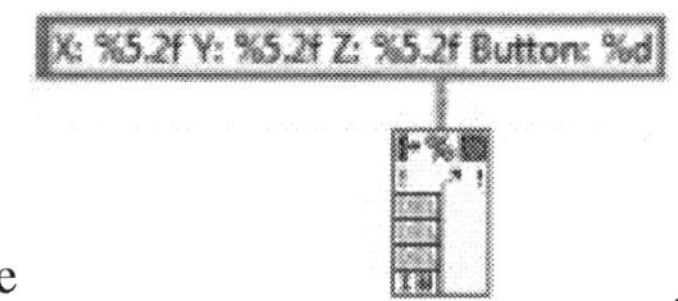

.

Step 9: In the Block Diagram window, right click on the screen. Under the Functions dialog box, click on Programming and then select Timing. Now click on Wait (ms) as shown in Figure 5.11.

Step 10: In the Block Diagram window, right click on the screen. Under the Functions dialog box, click on Programming and then select Dialog & User Interface. Now click on Simple Error Handler.vi as shown in Figure 5.12.

Step 11: In the Block Diagram window, right click on the screen. Under the Functions dialog box, click on myRIO and then select Button as shown in Figure 5.51.

Step 12: A Configure Button (Default Personality) dialog box appears as shown in Figure 5.52. Click on OK. Now place it in the block diagram and extend the button by dragging the bottom edge downwards such that error in and error out appear. It will look like 

.

Step 13: In the Block Diagram window, right click on the screen. Under the Functions dialog box, click on myRIO and then select Accelerometer as shown in Figure 5.53.

Step 14: A Configure Accelerometer (Default Personality) dialog box appears as shown in Figure 5.54. Click on OK. Now place it in the block diagram and extend the Accelerometer by dragging the bottom edge downwards such that error in and error out appear. It will look like 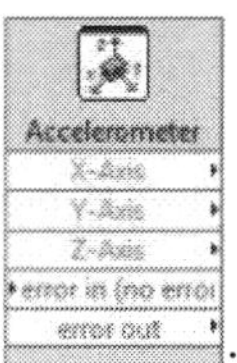

.

Step 15: In the Block Diagram window, right click on the screen. Under the Functions dialog box, click on myRIO and then select I2C as shown in Figure 5.62.

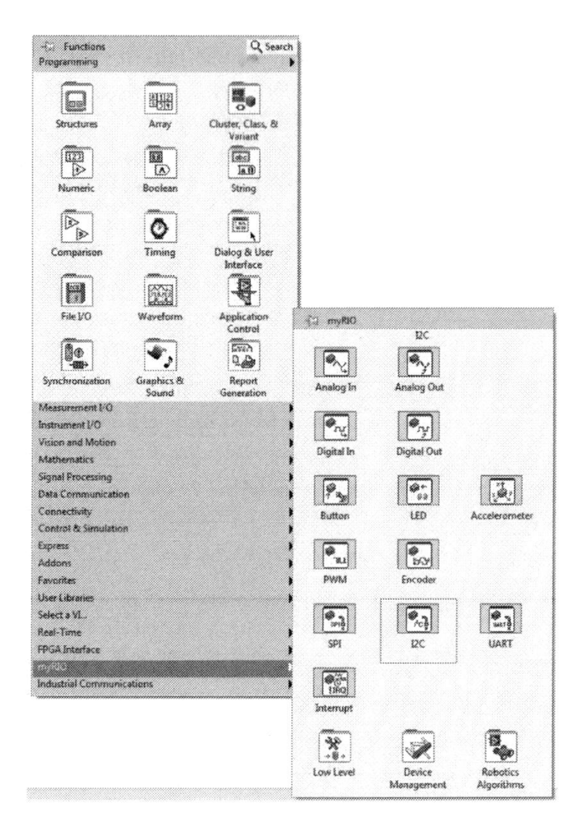

FIGURE 5.62 I2C.

Step 16: A Configure I2C (Default Personality) dialog box appears as shown in Figure 5.63. Click on OK. Now place it in the block diagram and extend the I2C by dragging the bottom edge downwards such that error in and error out appear. It will look like .

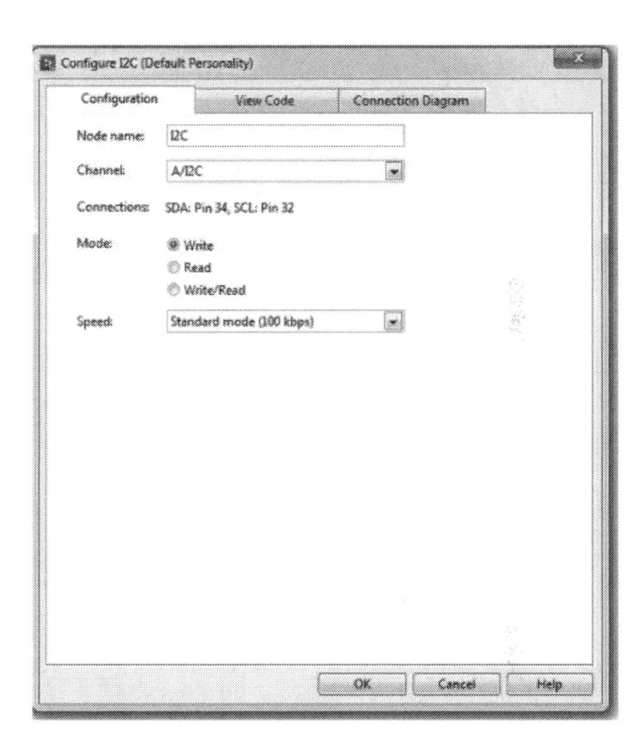

FIGURE 5.63 Configure I2C (Default Personality).

Step 17: Repeat Step 16 and place another I2C in the block diagram.

Step 18: In the Block Diagram window, right click on the screen. Under the Functions dialog box, click on myRIO and then Device Management. Now select Reset as shown in Figure 5.19.

Step 19: In the Front Panel window, right click on the screen. Under the Controls dialog box, click on Silver and then Boolean. Now select Stop Button (Silver) as shown in Figure 5.21. Rename the label [ESC].

Step 20: In the Front Panel window, right click on the screen. Under the Controls dialog box, click on Silver and then String & Path. Now select String Indicator (Silver) as shown in Figure 5.57. Rename the label LCD display.

Step 21: Connect all the blocks as shown in the final view of the block diagram window in Figure 5.64.

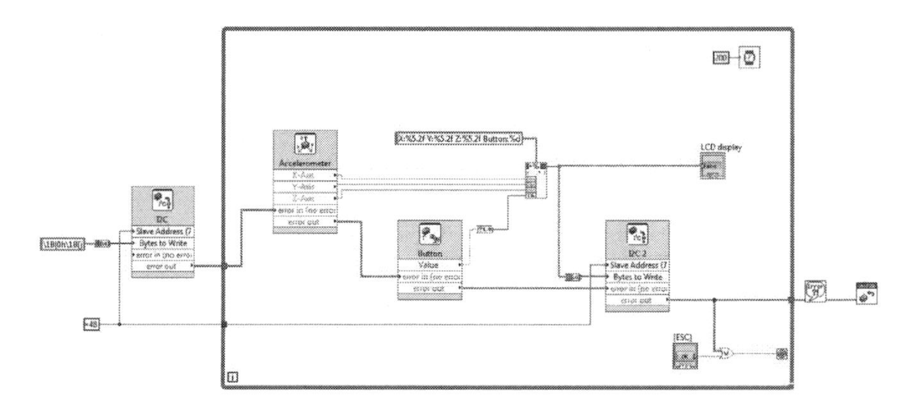

FIGURE 5.64 Final view of I2C based LCD Block Diagram.

Step 22: The final view of the front panel window is shown in Figure 5.65.

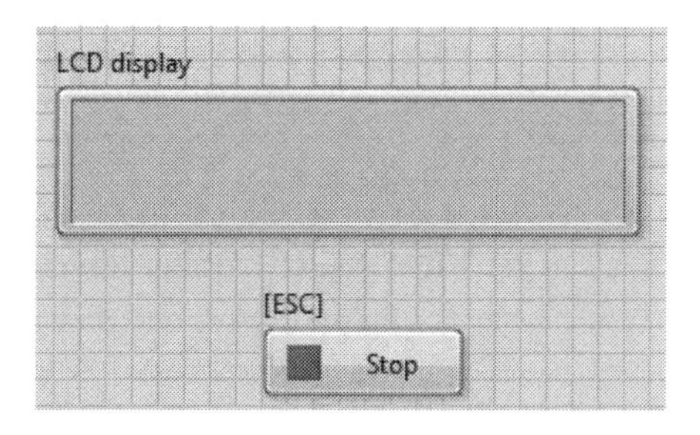

FIGURE 5.65 Final view of Front Panel I2C based LCD.

5.7 SPI Interface

SPI is the synchronous Serial Peripheral Interface bus most commonly used to send data between microcontrollers and small peripherals such as sensors, shift registers and LCDs. It uses separate clock and data lines, along with a select line to choose the device. Here we have interface LCD through SPI. Use a Digilent PmodCLS LCD to perform this experiment along with 4 Jumper Wires (F-F) and 4 Jumpers. Connect the Jumpers as shown in Figure 5.66. Connect J1/V of the LCD to pin 33 (B/+3.3 V) of my RIO using a Jumper Wire, connect J1/G of the LCD to pin 30 (B/GND) of my RIO using a Jumper Wire, connect J1/SI of the LCD to pin 25 (B/SPI.MOSI) of my RIO using a Jumper Wire and connect J1/CK of the LCD to pin 21 (B/SPI.CLK) of my RIO using a Jumper Wire.

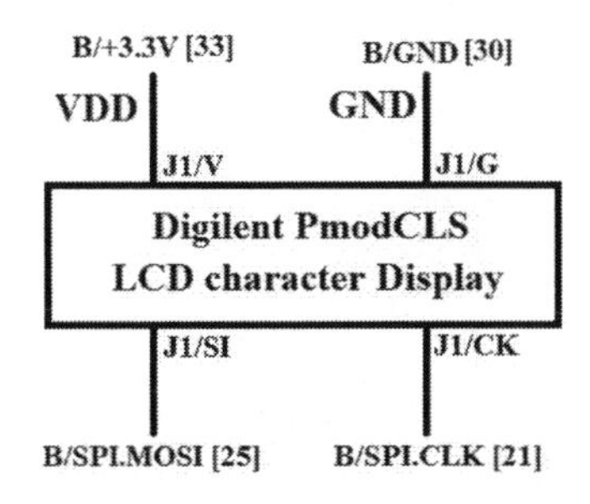

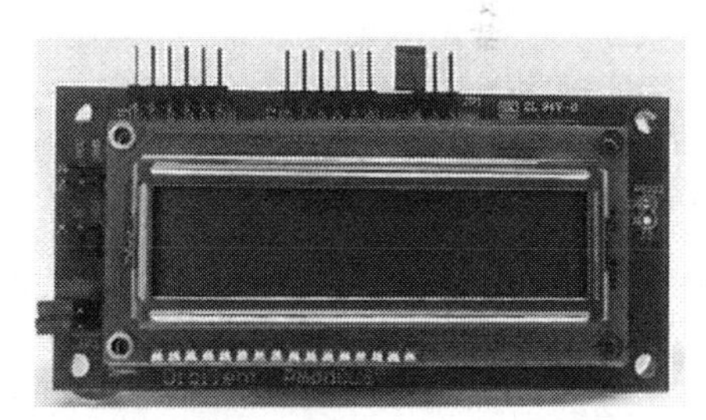

FIGURE 5.66 SPI interface.

Implementation Steps

SPI based LCD Interface

Step 1: In the Block Diagram window, right click on the screen. Under the Functions dialog box, click on Programming and then select Structures followed by the While Loop as shown in Figure 5.3. Now click and drag from the left top corner to the bottom right corner to form a rectangle.

Step 2: In the Block Diagram window, right click on the screen. Under the Functions dialog box, click on Programming and then select Numeric and then Conversion. Now select To Unsigned Word Integer as shown in Figure 5.67.

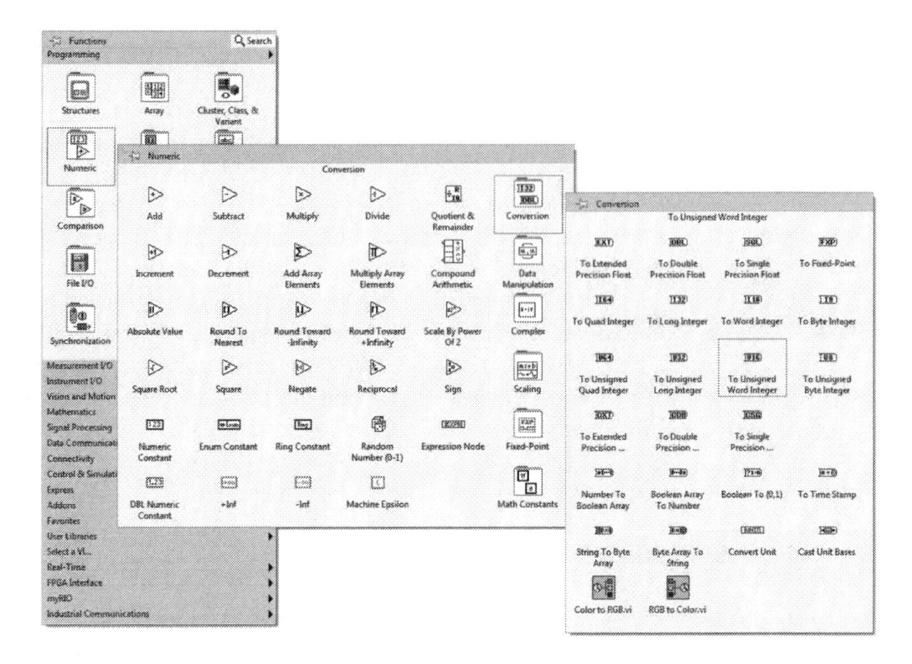

FIGURE 5.67 To Unsigned Word Integer.

Step 3: In the Block Diagram window, right click on the screen. Under the Functions dialog box, click on Programming and then select Numeric and then Conversion. Now select String to Byte Array as shown in Figure 5.61.

Step 4: Repeat Step 3 and place another String to Byte Array in the Block Diagram. Right click on the String to Byte Array and select Create and then Constant. Type in "[0h[j". Now right click on the constant and select "\" Codes Display. It will now look like "\1B[0h\1B[j".

Step 5: In the Block Diagram window, right click on the screen. Under the Functions dialog box, click on Programming and then select Numeric and then Numeric Constant as shown in Figure 5.7. Change the value to 200.

Step 6: In the Block Diagram window, right click on the screen. Under the Functions dialog box, click on Programming and then select Boolean. Now click on Or as shown in Figure 5.8.

Step 7: In the Block Diagram window, right click on the screen. Under the Functions dialog box, click on Programming and then select Boolean. Now click on Bool to (0,1) as shown in Figure 5.48.

Step 8: In the Block Diagram window, right click on the screen. Under the Functions dialog box, click on Programming and then select String. Now click on Format Into String as shown in Figure 5.49.

Step 9: Right click on Format Into String and click on Properties. Edit Format String appears as below. Click on Add New Operation four times. Type in "X: %5.2f Y: %5.2f Z: %5.2f Button: %d" in the Corresponding format string box as shown in Figure 5.50. Click on OK. It will look like 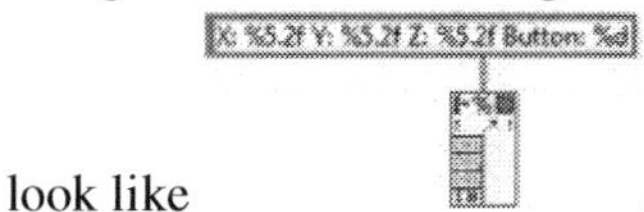 .

Step 10: In the Block Diagram window, right click on the screen. Under the Functions dialog box, click on Programming and then select Timing. Now click on Wait (ms) as shown in Figure 5.11.

Step 11: In the Block Diagram window, right click on the screen. Under the Functions dialog box, click on Programming and then select Dialog & User Interface. Now click on Simple Error Handler.vi as shown in Figure 5.12.

Step 12: In the Block Diagram window, right click on the screen. Under the Functions dialog box, click on myRIO and then select Button as shown in Figure 5.51.

Step 13: A Configure Button (Default Personality) dialog box appears as shown in Figure 5.52. Click on OK. Now place it in the block diagram and extend the button by dragging the bottom edge downwards such that error in and error out appear. It will look like 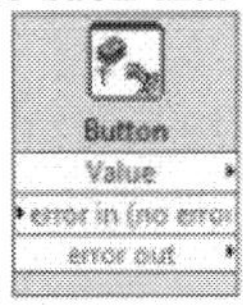

.

Step 14: In the Block Diagram window, right click on the screen. Under the Functions dialog box, click on myRIO and then select Accelerometer as shown in Figure 5.53.

Step 15: A Configure Accelerometer (Default Personality) dialog box appears as shown in Figure 5.54. Click on OK.

Now place it in the block diagram and extend the accelerometer by dragging the bottom edge downwards such that error in and error out appear. It will look like .

Step 16: In the Block Diagram window, right click on the screen. Under the Functions dialog box, click on myRIO and then select SPI as shown in Figure 5.68.

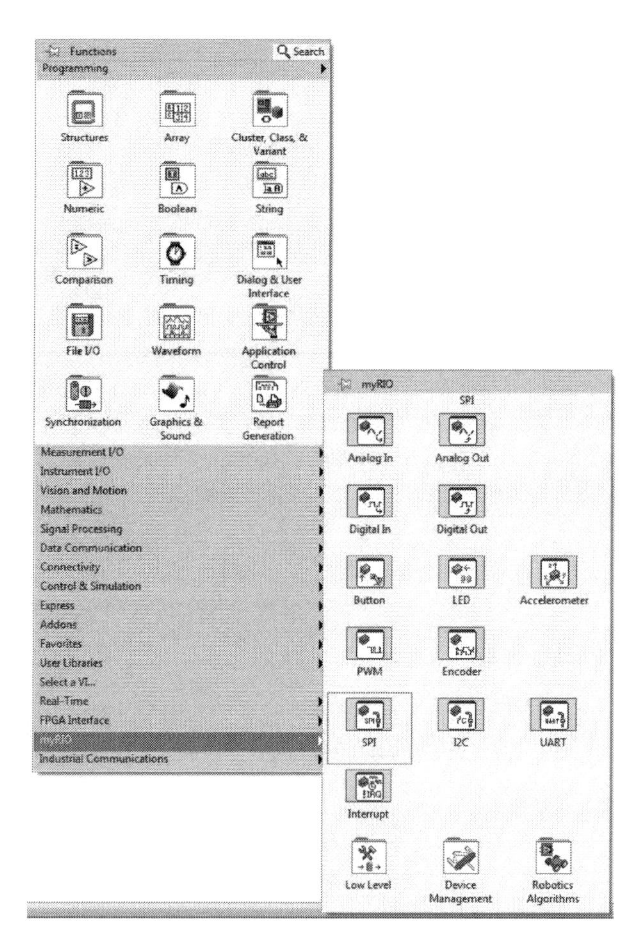

FIGURE 5.68 SPI.

Step 17: A Configure SPI (Default Personality) dialog box appears as shown in Figure 5.69. Click on OK. Now place it in the block diagram and extend the SPI by dragging the bottom edge downwards such that error in and error out appear. It will look like .

FIGURE 5.69 Configure SPI (Default Personality).

Step 18: Repeat Step 17 and place another SPI in the block diagram.

Step 19: In the Block Diagram window, right click on the screen. Under the Functions dialog box, click on myRIO and then Device Management. Now select Reset as shown in Figure 5.19.

Step 20: In the Front Panel window, right click on the screen. Under the Controls dialog box, click on Silver and then Boolean. Now select Stop Button (Silver) as shown in Figure 5.21. Rename the label [ESC].

Step 21: In the Front Panel window, right click on the screen. Under the Controls dialog box, click on Silver and then

String & Path. Now select String Indicator (Silver) as shown in Figure 5.57. Rename the label LCD display.

Step 22: Connect all the blocks as shown in the final view of the block diagram window in Figure 5.70.

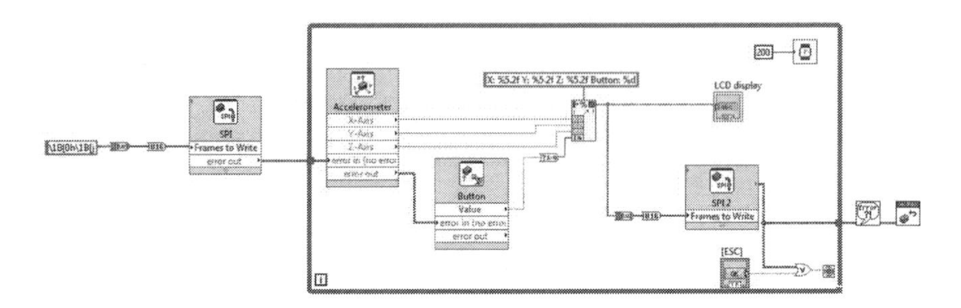

FIGURE 5.70 Final view of SPI based LCD Block Diagram.

Step 23: The final view of the front panel window is shown in Figure 5.71.

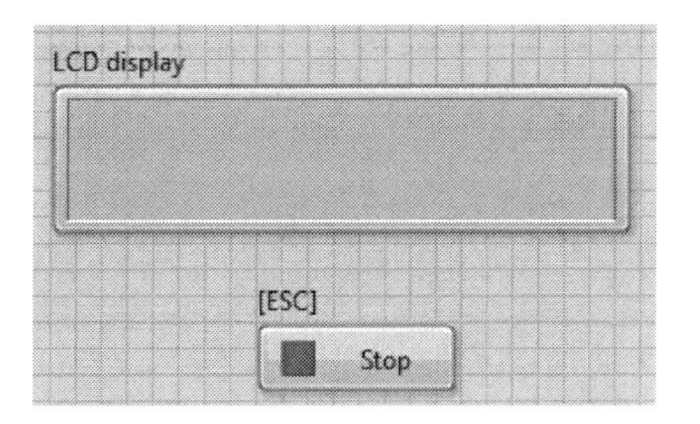

FIGURE 5.71 Final view of Front Panel SPI based LCD.

References

1. E. Doering. *NI myRIO Project Essentials Guide.* National Technology and Science Press: Austin, TX, 2013.
2. Y. Angal and A. Gade. LabVIEW controlled robot for object handling using NI myRIO, *IEEE International Conference on Advances in Electronics, Communication and Computer Technology (ICAECCT)*, IEEE, December 2–3, 2016.
3. M. Narvida. Learn LabVIEW FPGA on NI myRIO – Hello World! *All About Circuits*, February 2016.

Index